A NOTE ON THE AUTHOR

Chris Woodford had his first national magazine article published at the age of 13, and has been writing about science and technology pretty much ever since. After graduating from Cambridge University, where he specialised in physics and neuroscience, he trained as an advertising copywriter. Quickly concluding that junk mail and crisp packets did little to improve the world, he spent the next decade working as an environmental campaigner. He then moved into educational publishing and has written numerous popular science books, including the worldwide best-selling *Cool Stuff* series.

Chris also runs an educational website, Explain that Stuff, which demystifies everyday science for around half a million readers each month. He now lives in Dorset, surrounded by sheep and the sea.

Also available in the Bloomsbury Sigma series:

ATOMS UNDER THE FLOORBOARDS

FLOORBOARDS

*The Surprising Science Hidden
in Your Home*

Chris Woodford

BLOOMSBURY
sigma

Bloomsbury Sigma
An imprint of Bloomsbury Publishing Plc

50 Bedford Square   1385 Broadway
London   New York
WC1B 3DP   NY 10018
UK   USA

www.bloomsbury.com

First published 2015
Paperback edition 2016

British Library Cataloguing-in-Publication Data
A catalogue record for this book is available from the British Library.

Library of Congress Cataloguing-in-Publication data has been applied for.

ISBN (hardback) 978-1-4729-1222-0
ISBN (trade paperback) 978-1-4729-1822-2
ISBN (paperback) 978-1-4729-1223-7
ISBN (ebook) 978-1-4729-1224-4

2 4 6 8 10 9 7 5 3 1

Typeset by Deanta Global Publishing Services, Chennai, India

Printed and bound in Great Britain by CPI Group (UK) Ltd, Croydon CR0 4YY

Bloomsbury Sigma, Book Three

Contents

Introduction

How much would you say you have in common with science superstar Albert Einstein, the German-born genius who spawned Flash Gordon ideas like nuclear bombs and solar power? Load the question like that, and 'Not so much' is probably your first thought. And yet you might be surprised. To start with you share 99.9 per cent of Einstein's DNA (you share 99 per cent with a chimpanzee and 50 per cent with a banana, but we'll set these inconvenient truths aside for now). Did you hate school? Einstein did too: although a superb student, he dropped out of high school and picked up his education a year later at a technical college. Ever flunked an important exam? You and Einstein both: despite demonstrating brilliance in maths and physics, he botched his entrance exams to Zurich Polytechnic and had to put his studies on hold once more. Ever struggled to get the job you wanted? Einstein would have sympathised: after graduating he fired off endless, useless applications for every academic post he could think of, even considering a tiresome job in insurance to support his family, before finally winding up as a humdrum patent clerk. The most extraordinary scientist of the 20th century was, in many respects, a most ordinary sort of failure[1].

We love and revere Einstein not because he was brilliant beyond measure, but because he was a reassuringly flawed, human sort of genius, smart enough to know *everything* – and wise enough to know *nothing*. With squeaky, cheeky scribbles on the blackboard, he reinvented matter, energy, light and gravity – the most fundamental scientific concepts – in a baffling new theory called relativity that could stretch space and time like elastic. Yet, at the same time, he understood that science is, fundamentally, just another way of looking at things, and that scientific ideas, divorced from the sparkle and drudgery of everyday life, mean little or nothing to most. 'Gravitation,' he once wisely said, 'is not responsible for people falling in love.'

Science was Einstein's life, but it's probably not yours. You can pass three score years and ten without thinking about science for a moment – but you can't survive even a nanosecond without *using* science in one way or another. From Wi-Fi Internet to heat-reflecting windows, from brain scans to test-tube babies, science fires the technology that makes modern lives worth living, but still baffles most of us, even when we've spent years studying it in school. Recent opinion surveys show that 80–90 per cent of us are 'interested' or 'very interested' in science, and happily acknowledge how important it is. Yet 30–60 per cent think it's too specialised or too hard to understand, and two-thirds of 14-year-olds find it uninspiring. We confuse the ozone layer and climate change, we think nuclear power is more risky than crossing the road, and despite 70 per cent of us thinking that newspapers and TV sensationalise science, 86 per cent of us rely on precisely these unreliable media to keep us informed[2].

This book is designed to help put things right by exploring the science of everyday life in an engaging and entertaining way. Wandering around your home, it picks out the fascinating and surprising scientific explanations behind all kinds of everyday things, from gurgling drains and squeaky floorboards to rubbery custard and shiny shoes.

A note about notes; this book is meant to be an easy read for non-scientists, so I've kept detailed explanations, quibbles and qualifications to a minimum and avoided maths wherever possible. You'll find small superscript numbers dotted through the text pointing to a short section of notes and references at the end of the book.

Why is falling off a ladder as dangerous as being bitten by a crocodile? Is it better to build skyscrapers like wobbly jellies or like stacks of chocolate biscuits? How many atoms would you have to split to power a light bulb? Is there a scientifically correct way to stir your tea? There's nothing complex or baffling about the answers; it's not rocket science (even when it is). There's very little maths in this book to baffle or bore you; you really don't have to be Einstein to understand these things clearly and completely.

I'm no Einstein, but I've read some of his original papers and scratched my head over his elegant equations. To my mind, the deepest and truest thing he ever said was a simple sentence anyone can understand: 'The whole of science is nothing more than a refinement of everyday thinking.' That's the 'everyday thought' behind this book: it's science, if you will, for the rest of us.

CHAPTER ONE
Firm Foundations

In this chapter, we discover

> *How much a house weighs – and why it doesn't sink into the ground.*
> *Why your ankles work almost as hard as a building's foundations.*
> *How tall buildings stay upright – and what makes them fall down.*
> *Why soaring skyscrapers need to wobble like jellies.*

Scribbling novels, daubing portraits, rattling a piano through a Beethoven sonata – there are many things that are uniquely human. However, building things isn't one of them. There's no disputing that architects are among the most visionary of creatives, but the essence of what they do – dreaming up shelters that gravity can't tear down – is something many other animals do too. From snow domes built by grizzly bears to beaver dams that block the rush of the river, building for shelter is a trait shared by almost every species on land.

What makes humans different is the daring variety of the things we build. We have skyscrapers a quarter of a mile high and warehouses big enough to birth space rockets. We've got stone pyramids that have stood, silent and majestic, for about 5,000 years. We've all of us, at some point or another, built a house of cards that collapsed in seconds. We've got office blocks where 50,000 people simultaneously poke at keyboards or tell jokes around the water cooler. And we have phone boxes small enough to share the biggest secrets. Yet, for all their creative cunning, their desperate longing for originality, the boxes we plot and plan and raise so carefully from the ground are fundamentally no different from the homes animals scrape together from twigs and mud. That's because all buildings are shelters and all shelters, human or animal, have one thing in common: they use solid *science* to beat forces like gravity, wind, earthquakes and decay.

Safe as houses?

You probably don't trust everyone you meet, yet you place your faith in every building you set foot in: the first thought that crosses your mind on opening a door is seldom 'I wonder if this place is going to collapse?' People *sometimes* inspire confidence, but buildings always do. Not many humans push past the age of 90, but the oldest buildings in the world have been here 100 times longer. We say things are as 'safe as houses' confident in the knowledge that nothing much on Earth is any safer. Dreaming of falling is a common nightmare, but unless you live on a slippery cliff or in an earthquake zone in the Pacific, you'll wake up – thanks to a remarkable building – exactly where you nodded off.

Our confidence in buildings might be as sturdy as steel, underpinned by some pretty impressive scientific foundations, but the utterly static nature of homes, office blocks and other structures is extremely deceptive. Behind the sureness of their certain poise, buildings undergo a constant, invisible tug of war with forces like gravity, wind and earth tremors. Static buildings are in dynamic equilibrium. Mostly, it's a matter of deadlock: buildings go nowhere fast (or slow) because the forces trying to overthrow them, and those that try to keep them in place, are in absolute, immaculate balance. Yet the stakes are much higher than we ever bother to think about – something that only really becomes apparent on those rare occasions when giant structures tumble to the ground. How many of us, zooming up in the lifts of office blocks each morning, ever stop to think about the millions of tonnes of steel, glass and concrete looming over our heads, and what would happen if it all thundered down on top of us? The fact that buildings collapse so rarely shows that our quiet faith in science is very well placed.

How hard do buildings work?

To get a sense of how impressive buildings are, we need to understand the forces they have to cope with. Let's do that by calculating the force our own body has to withstand, then seeing how a typical house compares.

The weight on your feet

Force, to be clear, is the scientific name for a pushing or pulling action in a specific direction. Booting a football, heaving a bag of potatoes into your car, biting through a Snickers bar and hammering nails into a wall are some typical, everyday forces. The force that really dominates our lives – the one none of us can ever escape – is gravity: the pull between the Earth's hefty mass (a scale-busting 6,000,000,000,000,000,000,000,000 kg, or 6 trillion trillion kilograms in everyday English) and everything else that comes anywhere near. Gravity is the force that powers weight, the diet-driving, everyday heaviness we mostly manage to ignore. If you're a typical adult man, your weight is probably around 75 kg (165 lb), which probably doesn't even make you blink an eye until you think what it would be like to carry that weight around in your arms all day long. Racing up and down stairs, running, jumping, salsa dancing – whatever you're doing, you're doing with the equivalent of 75 bags of sugar stacked on your ankles.

That probably sounds quite horrific until you sit down and do the sums – take the mental weight off your feet, so to speak. The thickness of your ankles is obviously going to make a big difference: a couple of hefty tree trunks will spread your weight more effectively than a pair of pencils. I've just measured my own ankles and found that they're about 22 cm (9 in) all the way around, which means that the area of each one of my legs (the cross-sectional, circular area I'd see if I sliced through and looked down) is about 40 sq cm (6 sq in). Let's conveniently ignore the structure of the human body – the tissue, the bone and the way it's gift-wrapped in skin – and assume that legs are solid rods, much like tree trunks. If I weigh 75 kg, the **pressure** on my two ankles, together, is the force acting on them divided by the area it's spread across, which works out roughly the same as normal atmospheric pressure (the pressure of the air around us). Writing that in a more familiar way, it's about 14 pounds per square inch, which is about half the pressure in a typical car tyre, or – even easier to visualise – seven bags of sugar pressing on a postage stamp (next time you see someone waddling around on swollen ankles, you'll know why). The pressure on each one of my ankles is obviously only half this, which doesn't sound too bad, but it all depends what I'm standing on. Feet spread our weight over two or three times the area and reduce the pressure on whatever's underneath. Concrete or tarmac can easily cope with heavy humans; soft snow or wet sand on a beach will squash down a centimetre or so, leaving funky footprints that are fun to look back on, while feet will quickly sink deep into thick mud.

How does a building compare?
Houses, of course, don't have ankles: you don't have the entire weight of a building and its contents balanced on two thin columns at the base. Most houses (indeed most buildings) are built as straight columns perpendicular to the ground, so the cross-sectional area is roughly the same all the way up. Skyscrapers such as the Empire State Building are often tapered for extra stability: simplified ziggurats, they have somewhat wider bases gradually narrowing to the top. How do the numbers compare for the Empire State Building? You might expect a 102-storey building (about 380 m/1,250 ft tall) to exert a thumping great force on the ground – and it certainly does.

Just as in the case of our own bodies, however, what matters is not the force but the *pressure*: the area over which the force is spread. The base of the Empire State Building covers an area of roughly 8,000 sq m (86,000 sq ft) and the entire building weighs an estimated 330,000 tonnes (365,000 tons). This is as much as 4.5 million people, or the entire population of Calcutta, India[3]. Remarkably, thanks to the vast area of the base, that enormous weight produces a pressure on the ground of only four times atmospheric pressure[4]. We have to make a correction here, however, because buildings aren't solid blocks supported by their total ground area. At their simplest they're mostly emptiness resting on walls around the edges. Let's not quibble over the complexity of how a building is supported. Let's simply guess that 10 per cent of a building's footprint is wall and the rest is empty space, so we have to increase the pressure by a factor of about 10, which would make 40 times atmospheric pressure[5]. That sounds fairly hefty – which isn't surprising, given that we're talking about one of the world's biggest, tallest and heaviest buildings.

How does that compare with the force from a house? We need to know, of course, how much a house weighs, which isn't easy to guess. Rifling back through the archives of *Popular Mechanics* magazine, I found an article from 1956 in which the weight of a house had been estimated at about 122 tonnes (135 tons)[6]. A few years ago *Seattle Times* columnist Darrell Hay estimated that a typical house would weigh about 160 tonnes (160,000 kg/353,000 lb)[7]. Let's err on the side of plenty and guess the weight increases to about 200 tonnes (200,000 kg/441,000 lb), if we include all the junk we've loaded inside. That's hefty stuff indeed: an adult elephant might weigh about 5–7 tonnes, so we're talking about 30–40 elephants' worth of weight squashing down on your house. Now if the floor area of your home is a simple square of 10 m (33 ft) and we adjust, as before, to take account of the fact that the walls are doing most of the work, we find that a house is exerting twice atmospheric pressure or about 30 pounds per square inch on the ground – and about twice the pressure your legs have to support. So, remarkably, your thin and puny ankles have to withstand about half as much pressure as the walls of your home.

Why don't buildings sink into the ground?

This is the sort of question young children nag adults with all the time. Strangely, most of us are happy with an answer such as 'because they have foundations', even though we couldn't sneak such a response past an average seven-year-old, who might then say, 'But why don't the foundations sink?' The fact that buildings stay precisely where they are illustrates something very important about forces: they come in two distinct flavours. There are **static forces** (ones that explain why buildings and bridges go nowhere) and **dynamic forces** (those that tell us why skateboards and rockets do indeed go somewhere). The person who properly figured out the difference between the two was a moody and mercurial English mathematician everyone has heard of and some people insist is the greatest scientist of all time: Sir Isaac Newton.

It's the law!

Of the many contributions Newton made to modern physics, his biggest and best ideas are all about forces – and the simplest are summed up in three quick statements called the **Laws of Motion**. The first of these (Newton's first law) explains that an object stays exactly where it is unless a force acts on it. It's worth remembering this when you lose your car keys or glasses: things don't move by themselves. The second of Newton's laws tells us what happens when forces get together with objects to make them move: boot a football (apply a force to it) and it flies through the air. The final law is, in many ways, the most interesting of all. It says that when a force acts on an object, there's always another force (called a reaction) that's exactly the same size and acting in the opposite direction. Often you see this written in a rather terse form: 'action and reaction are equal and opposite'[8].

How does this relate to buildings? In the case of a skyscraper standing tall in the middle of a city, going precisely nowhere, Newton's first and second laws seem to suggest that there's no force acting on it. The first law states that things which aren't moving stay where they are when forces don't act on them, while the second law states that movement is kicked off by forces. Together, these laws suggest that a typical building is stationary because it's apparently force free. Yet we know that gravity – Earth's uber force – acts on a building all the time. Simplistically speaking, if Newton was right, buildings should grind downwards into the Earth and keep on moving indefinitely, or at least until they melt or sizzle up in the bubbling soup of the planet's core.

So why doesn't this happen? When the force of gravity pulls a building down into the ground, the ground pushes up against the building with exactly the same force. The two forces cancel out and the building goes nowhere. Why don't foundations sink into the ground? Because the ground pushes up on them. Even the tallest, heaviest skyscrapers seldom sink far. Many are supported on piles driven deep underground. The piles either rest on bedrock or, because they sink so deeply, are held in place by the friction between their own rough surface and the Earth or rock that surrounds them. We get motion only when forces are out of balance. If a building starts to sink into soft ground, the scientific explanation for this is that the ground can't produce enough upward force to balance the building's weight. The excess force acting down (the difference between the building's weight and whatever force the ground can muster) produces the motion that makes the building sink.

Where do forces come from?
If foundations prevent buildings from falling down, and forces stop foundations falling down, where do forces come from? Everything in the world around us is built from about

100 different types of **atom** – the 'Lego' blocks of life, better known as the chemical elements (things like iron, silver, carbon and oxygen). Groups of atoms join to make bigger structures called **molecules**, so two hydrogen atoms and an oxygen atom make a molecule of water (H_2O). Most of the forces we encounter every day originate inside atoms, between atoms, inside molecules or between molecules. There's more about atoms and molecules in later chapters, but for now let's just quickly consider how they might produce forces in buildings.

How atoms fight back
Suppose you build your house on a huge slab of iron. Iron is made from metal atoms packed tightly in a very orderly arrangement just the same as you would get if you dropped hundreds of identical marbles in a box. Each atom is like one marble. Atoms are mostly empty space but, like the most interesting sweets, they have layers that reveal themselves as you 'bite' towards the centre. Around the edge there are 'soft' clouds of electrons, which are negatively charged (like the negative, bottom end of a battery). In the centre there are protons and neutrons packed together in a hard inner core called the nucleus, and that's positively charged (like the positive, top end of a battery). The negative and positive parts of atoms are what prevent one atom from getting too close to another. You can't squeeze an iron bar very much because the negative electron clouds around the edge of one iron atom can't be pushed too close to the negative electron clouds around the next one. Like charges repel, just like the north poles of two magnets. The closer you push two atoms, the harder it gets to push them any closer.

The atoms beneath your feet
If you build a house on top of an iron slab you'll squash the atoms together a little bit, but only until they are so

close together that they won't be pushed any more. At that point the weight of your house pushing down is exactly balanced by the repulsive force between the atoms pushing back up. We don't build houses on slabs of iron, but the same principle applies to rock, soil and anything else. You can compact soil because it's crumbly stuff with a lot of air pockets inside it; you can squash sand because the gritty grains can slide past one another. Eventually, however, you'll reach a point where the ground is compacted and you can't squash it down any more. At that point you're trying to push atoms or molecules closer together than they actually want to go. Other types of force (the ones to do with electricity, magnetism and nuclear power) also originate inside atoms.

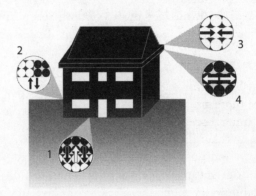

▲ **Four ways in which atoms help to stop a building from collapsing.** Clockwise from bottom left *1. Atoms in the foundations push back against atoms in the soil or bedrock. 2. Atoms in the side members of the foundations resist sliding past atoms in the ground, so the building is also partly anchored in the ground by friction. 3. The top of each horizontal beam is compressed slightly because of the weight it supports, but the atoms inside it resist being pushed too close together. 4. Similarly, the bottom of each beam is in tension, but the atoms are bonded together and resist being pulled too far apart.*

Every material compresses when you squeeze it with a force – even if the compression is a minuscule matter of just a few atomic diameters. Interestingly, that means a towering city office block is microscopically taller during the night, when it's empty, than during the day, when there are thousands of people, all adding their weight to the structure, all pushing it down. How much shorter is it? For a simple skyscraper about 400 m (1,300 ft) high, with about 50,000 average-weight people inside, the shrinkage worked out at about 1.5 mm (0.06 in)[9].

BRINGING THE HOUSE DOWN

Firm foundations hold your house high – but don't always keep it standing. Think of the simplest house you could build, from stacked stones or beach sand packed tight, and it's easy to see how forces hold the thing together. A house is essentially a mixture of materials, fused by gravity, which forces them inwards and down. In other words, what keeps a house standing are mostly forces of **compression** driving the walls down into the ground, with atoms in the ground pushing back up again.

You can't build a house from walls alone; you'll need plenty of cross-members – floorboards, roof joists and so on – stretching between them. These horizontal beams are in **tension** (where they bend apart at the bottom) and compression (where they squeeze tighter at the top), and transfer their own weight and any weight they carry to the walls, adding further to the forces of compaction that keep a house locked tight. Most homes stay standing for years, decades or even hundreds of years because most building materials – wood, stone and concrete – are incredibly strong in compression. This characteristic is well illustrated by Lego, a construction toy made up of interlocking plastic bricks. Lego is strong enough to support a tower 375,000 bricks high, stretching some 3.5 km (2.2 miles) into the sky. In force terms, each sturdy little brick can support a mass of 350 kg (770 lb) – about 4–5 times a person's weight[10]. Impressive stuff, you might think, until you happen to stand on one in your bare feet while you're tidying away the toys.

Signs of weakness

Houses fall down for many reasons, but ultimately only for one: something destructive creates a force bigger than the one that binds the structure together. Fires make homes unstable typically by burning through wooden roof beams or floor joists, or (at extremely high temperatures) weakening the steel bars inside reinforced concrete. The weight of the beams and the loads they carry then becomes too much for them to support, and the inside of the building collapses. Interestingly, the outer walls of a building are seldom directly damaged by fires, but are often knocked down by falling roof joists. One end of a joist usually falls before the other, so it swings down like a massive lever, its length amplifying the force it exerts when it crashes and smashes apart the walls beneath.

Although roofs usually fail first, the walls of a building can, sometimes, be the points of weakness. When wind hits a typical house, a sloping (pitched) roof helps it blow harmlessly past; the sloping, aerodynamic fairing on the cab of a truck serves the same function. Skyscrapers work a little differently: the bigger the face they present to a wind, the more high-powered air they collect. Some of this air will be bounced down to the ground, spinning around the streets at the base in vortices that can whip people off their feet; some will bounce even the tallest towers back and forth.

▷

Blown apart

Houses generally survive even the strongest winds although, in a hurricane, sustained pressure from outside can literally make a building implode. Many people in the Midwest of the United States cling to the mistaken belief that opening a window in a hurricane equalises the pressure inside and outside and reduces the risk of damage. The 'evidence' that this works is based on flawed logic and anything but science: if you open your windows and your house survives, that doesn't prove that the open windows were responsible. In fact, engineers have found that opening windows allows high-pressure, turbulent air to blast inside and increases the risk of the roof blowing off, making it more likely that the walls will collapse as well[11].

In a gas explosion pressure works in the opposite way. Most of us picture explosions as violent fireballs, but the galloping yellow flames are often incidental: explosions are magical, chemical reactions that can conjure up vast amounts of gas in a split second. Nitroglycerin, for example, is such a dangerous explosive because it readily turns from a liquid into a gas that takes up 3,000 times as much space. Semtex, the choice of terrorists, generates high-temperature gas at up to 30,000 km/h (18,000 mph) – about 30 times faster than a Jumbo Jet flies. When a gas leak or a bomb knocks down a house, it's as though a massive airbag has blown up inside it in the blink of an eye. That's what knocks the walls down, not the fire or heat of the explosion, which follow on behind.

How do everyday forces compare?

Art galleries, libraries and other civic buildings are often named after the wealthy people who made them possible. In much the same way, metric measurements typically take the names of the people who figured out the science behind

Table 1 Comparing force. *Forces are pushes and pulls that make objects move or, if they're perfectly balanced, do nothing at all. This table compares forces of different size.*

Source of force	Amount of force (newtons, N)
Weight of an apple	1
Weight of a bag of sugar	10
Force of bite from a person's jaw	~500–1,000
Force of a small car engine during overtaking	2,000
Force needed to break a human bone	3,000–5,000
Force of bite from an alligator	16,000
Force of a car crashing	50,000
Force powering a typical jet plane (four engines)	300,000
Weight of a typical house and contents	2 million
Total force from the Space Shuttle at blast-off	32.4 million

them. Isaac Newton's spectacular contribution to the science of forces is recognised in the most fitting way: in modern scientific speak, forces are measured in things called **newtons** (N).

What does a newton feel like? Newton reputedly got his inspiration for the law of gravity when an apple tumbled from a tree and plopped on his head (a story widely believed to be apocryphal). If an apple weighs about 100 g (3.5 oz), the force of gravity pulling down on it is about 1 newton. So, remembering what we said up above about forces balancing one another, if you balance an apple in the palm

of your hand, you have to push upwards with a force of
1 newton to keep it perfectly still.

We can convert kilogram weights into forces if we
multiply them by 10. That's because Earth's gravity pulls
every kilogram towards it with a force of 10 newtons. So if
you weigh 75 kg (165 lb), Earth pulls you down with a force
of 750 newtons. Simple. What about other everyday objects?
If your house weighs 200 tonnes (200,000 kg/440,000 lb),
that makes a force of 2 million newtons (2 meganewtons).
Table 1 shows some other forces for comparison.

WHY DON'T SKYSCRAPERS BLOW OVER?

If all houses have to worry about is sinking into the ground,
taller buildings have an added anxiety: not getting blown
down in the wind.

Big foot

In the real world skyscrapers astound us by soaring to the
sky. If you had to guess how much higher a skyscraper is
than it is wide, what would you say? Ten times? Fifteen
times? Twenty times? More? To most people's surprise, it
turns out that even the tallest buildings are seldom more
than about **seven** times higher than their width at the base.

The secret of a skyscraper is that we don't notice it has
big 'feet'. The Empire State Building is about 100 m
(328 ft) wide and 380 m (1,250 ft) high, so its height-to-
width ratio is a mere 4:1. Despite its height, the Eiffel
Tower manages a ratio of just 2.4:1 because its legs are so
far apart. Your own 'base width', standing legs apart, is
probably 30–50 cm (12–18 in), so as a 'tall building', you're
about 4–5 times higher than you're wide. A few skyscrapers
really do push the boundaries. A skinny residential block
in Hong Kong called Highcliff manages an astonishing

20:1 ratio. Once you reach that point, it takes engineering ingenuity – not just big feet – to keep a building upright.

Secrets of the windy city

One of the fascinating things about skyscrapers isn't that they stay precisely where we build them, but that they don't. For reasons we explore in Chapter 15, the speed of a wind increases dramatically the higher you climb from the ground. So a building that's sticking up 0.5 km (0.3 miles) and, at the same time, is broad enough to stay upright, is going to present quite an attractive target for a passing gale. Sunk motionless in the ground but with a hefty force (the wind) pounding horizontally at roof level, the whole building is going to work like a giant lever, and a big enough wind force could, theoretically at least, snap it in half or uproot it like a tree.

Now that might suggest that we should make buildings as stiff and rigid as possible, but it turns out that wobbly buildings are much more likely to survive. That's intuitively obvious if you think about what happens to a tall jelly when you shake it, compared to what would happen to a vertical stack of chocolate biscuits. Much like jellies, skyscrapers are designed to wibble and wobble, relatively slowly, in challenging winds. The Twin Towers, for example, wandered a rather frightening 1 m (3 ft or so) at the top, while Taipei 101 (a much more recent and not so well-known skyscraper in Taiwan) sways 60 cm (2 ft). By comparison, the (older and significantly shorter) Empire State Building's wobble is a much less frightening 8 cm (3 in) or so[12].

Damping down

What matters isn't just how much a building wobbles, but how quickly or (better still) slowly it does so. Skyscrapers swing back and forth in a predictable amount of time like

▷

pendulum clocks. Chicago's John Hancock Tower oscillates with a period of 8.3 seconds (about eight times slower than the tick of a clock)[13]. Faster oscillations than this make people air-sick inside. Skyscrapers don't fall over because wind and earthquakes make them sway like pendulums and, just like pendulums, their oscillations gradually die down until the energy is dissipated and they stand still and tall once more.

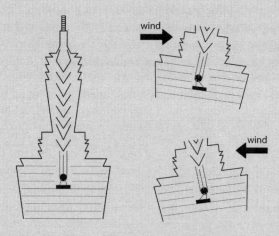

▲ **Preventing wobble – the Taipei 101 skyscraper.** *A 660-tonne (730-ton) ball called a tuned mass damper is used to help the 509-m (1,670-ft) high Taipei 101 skyscraper resist wind wobbles. The damper is fastened loosely near the top of the structure with hydraulic rams, similar to the shock absorbers in a car. As the wind thumps the building on the left or right (exaggerated in this diagram), the incredibly heavy damper does its best to stay where it is. That means that it has to tug on the hydraulic rams, damping down the vibrations of the building as do a car's shock absorbers, and stopping the high-flying occupants from getting motion sickness[14].*

CHAPTER TWO

Upstairs, Downstairs

In this chapter, we discover

>*Why you probably shouldn't eat a chocolate-chip cookie straight after climbing the Empire State Building.*
>*How much electricity you can get from a bolt of lightning.*
>*How long it would take to make a cup of coffee if you could get the energy from a hamster running around a wheel.*
>*Why falling off a ladder is as bad as being bitten by a crocodile.*

What do these things have in common: a red-hot, stinging slap across the face; the startled applause of pigeons scattering from your window ledge; the effervescent fizz of a hangover pill dropped into a glass of water; a bloodshot wink from the smoke alarm in the middle of the night; and a fly trapped, wrapped and cocooned in the fatal trampoline of a spider's web? They're all forms of energy, hiding in and around your home[15]. Invisible and inscrutable, energy is the ultimate riddle. Practically defying definition, it powers the very thought processes that try to understand it. Let's put some brain energy to good use – figuring out energy itself.

Weighing in
Are you a healthy, hearty 'stairs' person – or would you rather sneak up in a lift (elevator)? The fatter you are, the less likely you are to choose to struggle with steps. While most people would put that down to sheer, waddling laziness, it's actually a basic and very compelling bit of science. Consider Andy and Bob, a couple of imaginary engineers tasked with repairing a broken lift pulley on the highest floor of the

Empire State Building, 380 m (1,250 ft) above the Manhattan skyline. Andy weighs in at a hefty 95 kg (210 lb), while Bob, his tiny assistant, is a svelte and athletic 65 kg (140 lb). With the lift out of action, there's no option but to tread the unforgiving stairs – all 1,870... 1,871... 1,872 of them.

Now we all understand instinctively that going upstairs is hard work. The higher you go, the further you have to heave your body against Earth's gravitational pull and the more **energy** you use up in the process. What's less obvious is that fatter people have to work much harder than thinner ones. The bigger your body, the more **mass** (the scientific word for 'stuff' – what we loosely think of as weight) you heave around, and the more energy it takes to haul that mass upwards. Why is this fact less obvious? Because it never occurs to us to carry out a scientific experiment: it's not as though we can climb something, put on some weight, then climb it again 10 minutes later to compare the difference.

However, there is a difference – and it's quite startling. Heaving bulky bags of shopping upstairs is enough to give us a pretty strong clue to why this is.

How much more energy does Andy use on the climb to the top? It's a quick scribble of school science to work it out: amazingly, the answer is 120 kilojoules[16] (kJ, an amount we explain in more detail below), which is just about how much electricity you'll use boiling the water for a mug of coffee. That's how much harder Andy's body has to work on a long climb purely because of the extra weight he carries. When they finally reach the top, whacked out and panting for mercy, Andy digs deep into his kit bag for a pack of compensatory chocolate-chip cookies. 'I've earned it,' he says, smiling, and stuffing a couple of the energy discs deep in his face while tossing the packet over to Bob. His colleague, nosing through the small print on the wrapper, politely declines: two cookies contain 108 Calories (450 kJ). *Theoretically*, if all the food energy were used for moving Andy's body upwards (in other words, if 100 per cent of the energy in the food were converted into what we call mechanical energy), this would be more than enough to power a climb like the one they've just done. 'Next time,' he muses quietly to himself, 'Andy's probably going to find that walk even harder.'[17]

What is energy, anyway?

There's a sense in which we understand energy – and a sense in which we don't. We all know it's the invisible fuel that powers our lives. We know we have to pump petrol into the car to avoid breaking down. We know we have to gobble down food two or three times a day to keep our bodies ticking over. And we know we have to keep paying the gas bill if we want steamy showers and a cosy home. Yet how many of us could estimate, accurately, the total amount of electricity we used yesterday (at home, work or school), how much power a city the size of New York or New Delhi uses every year, or how many power stations we need to build over the next decade to keep the lights on? The basic concept of energy is easy enough to grasp. We understand it *qualitatively*; where we fall down is in understanding it *quantitatively*. That's why we're always fuming about our gas bills, why the world's constantly wobbling on the edge of an energy crisis and, quite probably, why a growing number of us are morbidly obese, eating so much more energy than we use that we put our lives at risk. Measuring energy is a great way to get a deeper understanding of what it is and how it powers our world – and that's what this chapter is all about.

Joule's joules

Business people swear by the mantra that 'you can't manage what you can't measure', and the same holds true for science – with a twist: in science, it's more a case of 'you can't *understand* what you can't measure'. The way to get to grips with scientific ideas like energy is to start putting numbers to them. That means we can compare the amount of energy we use doing everyday things with the amounts we can produce in various ways. Do we use more energy climbing the Empire State Building than we get back by nibbling on a cookie at the top? In theory, no (in reality, yes). Can we make enough electricity with a wind turbine to power an entire village? Yes. If we hook a bicycle dynamo to a kettle and pedal like mad, how long do we need to keep going to boil a big saucepan full of water? The answer is 21 hours, although a racing cyclist powering a decent electricity generator could manage it in as little as a quarter of an hour[18].

Scientists measure energy in units called **joules**, named after 19th-century English physicist James Prescott Joule, who made some of the first experimental measurements of energy (which he liked to refer to as *vis viva*, from the Latin 'living force')[19]. In science, one joule is one unit of energy – but what exactly is a joule? What does it look and feel like? We've already seen that it takes about 120 kilojoules (120,000 joules) to boil a mug full of water, so a single joule doesn't look as though it's going to be that impressive. Take an orange (which weighs roughly 100 g/3.5 oz) and lift it a metre, and you'll use about one joule of energy in the process. Doesn't sound much, does it? Now turn the comparison around. Boiling a mug of water (120,000 joules) is the same as lifting 120,000 oranges (12 tonnes of them) a metre in the air, or shooting one orange 120 km (75 miles) into the air (about 14 times the height of Mount Everest). That sounds more like it.

How much energy does it take?

From riding a bike to running a marathon, it's relatively easy to work out in theory how many joules of energy you need to do all kinds of everyday things (see Table 2). That's the 'debit', demand or consumption side of energy. In a similar way, it's reasonably easy to figure out how much energy is hiding inside a chocolate-chip cookie, a car battery or a lump of coal, and see what you could do with it. By comparison, that's the 'credit' or supply side of energy. Thanks to the work of James Joule, we know that the energy books always balance: the energy 'credits' and 'debits' match exactly.

Table 2 Comparing energy. *One way to get a feel for energy is to compare the amount we need to do different things with the amount we can produce in different ways. In this table we're comparing ways of making energy (italic) and using it (regular text). For each item in the table, I've calculated the amount of energy involved in joules (an unintuitive unit most of us are not that familiar with), kilowatt hours (the unit most of us are billed for energy by gas and electricity companies) and 'Empire State Building climbs' – the minimum amount of energy an average-sized, 75-kg (165-lb) person would need to use to climb to the top of the Empire State Building if they took the stairs[20]. From this table you can see that a lightning bolt contains the same amount of energy as you'd use if you climbed the Empire State Building about 17,000 times, while an hour's swimming is like climbing the building seven times.*

Energy	Joules	Kilowatt hours	Empire State climbs
Average lightning bolt	*5,000,000,000 (5 billion)*	*1,400*	*17,000*
Nuclear plant for 1 second	*2,500,000,000 (2.5 billion)*	*700*	*8,333*
Burning 1 litre of petrol	*36,000,000*	*10*	*120*
100 watt lamp for 24 hours	8,640,000	2.4	30
1 kilowatt hour	3,600,000	1	12

▷

Energy	Joules	Kilowatt hours	Empire State climbs
1 hour's vigorous swimming	2,000,000	0.6	7
Wind turbine for 1 second	1,000,000	0.3	3.3
Eating 2 hard-boiled eggs	670,000	0.19	2.2
Toaster for 3 minutes	540,000	0.15	1.8
Climb Empire State Building (lift 75kg up 400m)	300,000	0.08	1
AA alkaline battery	10,000	0.003	0.03
Home solar roof for 1 second	4000	0.001	0.013
Climb your stairs at home (lift 75kg a height of 3m)	2250	0.0006	0.008
10 watt lamp for 1 second	10	0.000003	0.00003
Lift an orange 1m	1	0.0000003	0.000003

What is power?

While it's interesting to think about different amounts of energy, it's not always that useful. Although it takes quite a bit of energy to puff your way up the Empire State Building, what really matters is how long you allow yourself to do it. Suppose you stroll up the stairs in a generous eight hours, pausing every so often to admire the view: that works out at about four steps per minute, which might be tedious, but it won't tax your body very much. Set yourself a more ambitious target of getting to the top in half an hour, and you'll be pounding and panting upwards at roughly one step a second, which is much more challenging. When it comes to using or producing energy, it's essential to factor in the time we have available.

We can do this by talking about **power**, which is the rate of using or making energy (or the amount of energy divided by the time it takes to use or make it). As in the case of energy, we can get a handle on power by measuring it – and we do that with units called joules per second, better known as watts. A 100-watt lamp uses 100 joules of energy per second. If Andy, our mythical 95-kg engineer, clambers to the top of the Empire State Building in half an hour without collapsing from a cardiac arrest, he'll be using energy at a rate of about 200 watts[21].

How much power does it take (or make)?
When it comes to producing energy, watts are tiny amounts of power. A hand crank can produce about 10 watts, but if you've ever tried using one to power something like a table lamp or a laptop computer (as I once did), you'll know it's physically very tiring[22]. Generating decent amounts of power is hard work. A fairly large wind turbine produces about 2 megawatts (2 million watts), which is the same as 200,000 hand cranks or 10,000 theoretical Andys simultaneously puffing their way up the Empire State Building. A very large coal or nuclear power plant thumps out about 2 gigawatts (2,000 million watts), which is the same as about 1,000 wind turbines. That's why so many wind turbines are suddenly springing up all over the place: it's not that people necessarily like building them, but that we need at least 1,000 of them to produce the same amount of electricity as a single large power station.

Comparing power

Power is energy divided by time, so if you want to find out how much *energy* something like a tumble dryer uses, you can simply multiply its power consumption rating (which might be 3,000 watts or 3,000 joules per second) by the length of time you use it. If it's on for an hour, that's 60 minutes made of 60 seconds, or 3,600 seconds. So the total energy you use is 3,000 joules per second × 3,600 seconds,

Table 3 Differences in power generation. *Power is the amount of energy something makes or uses in one second. It's hard to visualise how much power you can get out of a steam locomotive, never mind a power station or a space rocket. If, however, we relate each of these things to the power of a Porsche Turbo (a powerful sports car) or a hamster running around a wheel, it all starts to make more sense. That's not to say that you can directly replace any of these power sources with Porsche Turbos (which could never blast you into space, because their engines need oxygen) or hamsters (which would quickly tire out). While a nuclear power station can produce any amount of energy (by producing high power for decades on end), hamsters will only produce power for a few minutes before collapsing from exhaustion, so the total energy they can make is trivial[23].*

Power source	Watts	~Porsche Turbos	~Hamsters
Space Shuttle on take-off	11,000,000,000 (11 gigawatts)	28,000	22 billion
Hoover hydroelectric dam	2,000,000,000 (2 gigawatts)	5,100	4 billion
Nuclear power station	1,500,000,000 (1.5 gigawatts)	3,850	3 billion

Power source	Watts	~Porsche Turbos	~Hamsters
Steam locomotive	1,500,000 (1.5 megawatts)	4	3 million
Diesel truck engine	450,000 (450 kilowatts)	1.2	900,000
Porsche Turbo car engine	390,000 (390 kilowatts)	1	800,000
Microwave oven (full power)	1000 watts	0.003	2000
Hand-held vacuum cleaner	400 watts	0.001	800
Human hand crank	10 watts	0.00003	20
Bicycle dynamo	5 watts	~0	10
Hamster on wheel	0.5 watts	~0	1

or about 10 megajoules (10 million joules, which is the same as 10,000 kJ). In exactly the same way, we can figure out how hard it is to do everyday things. If it takes 120 kJ (120,000 joules) to boil water for a cup of coffee, and if it were feasible to do it using some kind of hamster-powered wheel, how long would it take? A hamster produces 0.5 watts or 0.5 joules in a second, so it would take 240,000 seconds, or about three days. Using a hand crank gives us 10 watts, so we could do it 20 times faster, in about three hours[24]. At the opposite extreme, what if you had your own personal nuclear power station? Hooked up to a suitable kettle, it would give you 1,500,000,000 joules in just one second, so you'd get your coffee in about a 10-thousandth of a second – instant coffee indeed.

The cost of energy

One thing that makes energy really confusing is the way we pay for it. Your gas or electricity bill is probably measured in units called kilowatt hours, which sound as though they might be units of power (because the word 'watt' is in there), but they are actually units of energy. Power is energy divided by time, so energy divided by time, then multiplied by time again, is simply energy. A kilowatt hour (1 kWh) is the amount of energy you use if you leave something with a power of 1 kilowatt (1 kW) running for a whole hour. How does this work in practice?

- A vacuum cleaner with a 1000-watt motor uses 1 kilowatt hour if you have it running continually for a whole hour.

- A fast electric kettle uses about 3 kilowatts, so if you had it boiling away constantly for 20 minutes, that would use exactly 1 kilowatt hour. It doesn't matter how much water there is inside it: that simply affects how quickly it comes to the boil.

- A low-energy light bulb nibbles through a measly 10 watts of power, so you could keep it on for 100 hours (about four days) before it used a single kilowatt hour.

- Andy, our one-stair-per-second, overweight Empire State Building climber, uses energy at the rate of 200 watts, so he could climb for a full five hours before clocking up a kilowatt hour.

Whatever job you want to do, it usually takes the same amount of energy to do it[25], but using more power results in you doing it faster. Whichever way you choose to get to the top of the Empire State Building, you always need to lift your body mass (which doesn't change) by the same distance, so (in theory) you always need the same amount of energy. If you go by lift, the electric motor moves your body up much more quickly than you can move it yourself, which is another way of saying that the motor is more powerful (it supplies the same energy in less time). However you boil a litre of water, you'll need to use 378,000 joules of energy (378 kJ) to do it. You can boil your water in an electric kettle, on a gas stove, over a camp fire or by stirring it with a spoon. All these things will boil the water eventually (even, if you did it just the right way, using the spoon[26]), but each supplies energy at a different rate, works at a different power, and takes more or less time. If you use a 3-kW fast-boil kettle, you'll boil your water three times faster than if you use a 1-kW travel kettle. You'll supply exactly the same amount of energy but three times quicker, so with three times the power. Either way, you'll use the same number of kilowatt hours and it'll cost you exactly the same amount.

Where does energy come from and go?

Money doesn't suddenly appear in your bank account or vanish from your wallet with no explanation. We earn and spend money in constant transactions with other people – and energy works in the same way. If you want energy, you have to 'earn' it from somewhere (by eating food or filling up your car's petrol tank). If you want to do something, you 'spend' some of your energy and end up with less than you had before. While it's possible to fiddle the financial system by printing or counterfeiting money – making it appear out of thin air – you can't pull off the same trick with energy, no matter how hard you try. There's a fixed amount of energy in the Universe and all we can do is 'trade' it in a zero-sum game: every energy gain somewhere is exactly matched by an energy loss somewhere else. This is an absolute and utterly fundamental law of physics and it goes by the name of the **Law of Conservation of Energy** (not to be confused with what we normally mean by 'energy conservation' or saving energy, which is something quite different).

James Joule's experiments helped to establish the law in
its best-known and most modern form: we can't create or
destroy energy, only convert it from one form into another.
Joule offered a colourful example, showing how the
temperature at the bottom of a waterfall would be higher
than that at the top because the tumbling water's energy
would turn back to heat as it smashed into the river below.
With a few quick sums, he found that Niagara Falls would
be a fifth of a degree warmer at the bottom than at the
top[27]. Unfortunately, his cunning attempt to test this
theory – by taking some super-sensitive thermometers
with him to measure waterfalls during his honeymoon in
Chamonix, France, in 1847 – didn't succeed. The water
sprayed about too much on impact, making it impossible
to measure the temperature increase with enough
accuracy[28].

How does all this work when you boil yourself a litre of
water? One way or another, you feed 378 kJ of energy into
the pan. It might be 378 kJ of electrical energy zapping
through a power cord into a kettle's heating element, or it
could be 378 kJ of gas roaring on a hob. In theory, if you
were careful to avoid heat losses from the pan, it could even
be 378 kJ of energy supplied by stirring the water with
super-rapid flicks of your wrist. In that case the energy
would come from your own body – maybe from chocolate-
chip cookies you gobbled down earlier. Let's say that's the
way you do it. By the time the water's boiling, your body
will have lost 378 kJ of energy and the water will have
gained an equal amount – in the form of heat. If you then
drink the hot water, you'll get some of that heat energy
back again: a hot drink warms you up so your body doesn't
have to work quite as hard to maintain its temperature.

Why does it hurt?
Understanding *amounts* of energy explains all kinds of
things that don't seem to have anything to do with

energy – such as why cars crumple when you crash them, why a burn from steam hurts so much more than one from boiling water and why it's dangerous to fall off a ladder. The answer in all three cases is because 'energy has to go somewhere' – the Law of Conservation of Energy, in other words.

Crashing a car

If you're haring down a motorway at 150 km/h (90 mph/ 40 m per second), in a sports car that weighs 1,500 kg (3,300 lb), the car has energy because it's moving. We call that **kinetic energy** and we can calculate it with a fairly simple maths formula: it's half times the mass, times the speed, times the speed again[29]. Plugging in the numbers, you'll find that the car has a little over 1 megajoule of energy. Once you've crashed, your speed is zero and you have no kinetic energy. So the act of crashing is equivalent to shedding 1 megajoule of energy in the blink of an eye – and the power involved is equal to the energy you have to lose divided by the number of seconds it takes you to crash. So if your car crumples in, say, half a second, you're getting rid of energy at a rate of 1 megajoule/0.5 seconds = 2 megawatts, which (we can see from Table 3 on page 34) is roughly the power of a steam locomotive. That's why a car crash is so violently dangerous, and why your car crumples. The longer it takes to crash, the smaller the force involved (on the car and on the people inside), and the more likely you are to survive.

If you walk away from a car crash unscathed, you can thank the fact that cars are deliberately designed to scrunch up so that they disperse kinetic energy more slowly, reduce the force on your body and save your life. The worse your car looks, the more grateful you should feel. You survive because your car doesn't: it's a basic design feature that it sacrifices *itself* to save *you*. In theory, we could design indestructible cars that would escape any kind of crash with barely a scratch. However, in that case there'd be nothing to soak up the energy in a collision, the forces on our body would be enormous and even a modest shunt could prove fatal[30].

Burning yourself with a kettle

Exactly the same reasoning applies to a kettle burn: the energy in the hot water has to go somewhere – and it goes straight to your fingers, cooking the living tissue. Steam at 100°C (212°F) contains vastly more heat energy than water at the same temperature. That's because you need to add extra energy to water to push apart the molecules inside it and turn it into a hot vapour (gas). Stick your hand in steam and you get two doses of heat. You absorb the heat as the steam turns back to boiling hot water, then you get the extra heat as the boiling water cools down to the temperature of your hand.

Falling off a ladder

Climbing up a ladder takes energy. Fall off a ladder and that energy (which is called **potential energy**, because it's notional, stored energy you can use in the future) has to go somewhere: straight into your body. If you climb a ladder onto your roof (10 m/about 30 ft), and you weigh about 75 kg, your potential energy is 7,500 joules[31]. Fall off a ladder so your head smashes onto concrete in an impact that lasts (say) a tenth of a second, and your body has to deal with a power of 75,000 watts (75 kW). The force on your head is directly related to the rate at which your body gets rid of the energy: the faster the energy goes, the more it hurts. Plug in the numbers and you'll find that what hits your head is approaching the sort of force a crocodile can apply with its jaws[32]. Plenty enough to break bones – and even kill you. Why does it hurt? Why might you die? Because energy has to go somewhere.

Upstairs and downstairs – in your own home

All this talk of joules and watts – and, more to the point, pounds and pence – should leave one thing beyond doubt: everything we do needs energy and, one way or another, costs money. That's the bad news about the Law of Conservation of Energy. The good news is that using energy isn't always as pointless as burning money to keep warm – we can sometimes get our energy back again.

If you climb upstairs at home, you'll convert food into energy – and the kind of cunning energy (potential energy) you can then use to do other things. If you're lucky (and eccentric) enough to have a fireman's pole at home, slide down it to the ground floor and you'll get your potential energy straight back. It'll be converted into kinetic energy (motion and speed) as you race to the ground. In theory, you could build some sort of human hamster wheel (a belt-pulley-gear contraption) that would spin around as you slid past, converting your kinetic energy into electricity. Since every step of the process would be a little bit inefficient (wasting energy making heat and noise), you wouldn't be able to generate as much electrical energy on the way down as the energy you spent in food on the way up – but you'd still salvage something. Of course, what you can't do is get your highly calorific food back. Although the Law of Conservation of Energy permits it, we've yet to invent the mechanism that can re-present you with a fully formed chocolate-chip cookie as you arrive back on the ground floor of the Empire State Building, having previously puffed your way to the top. That's an important lesson to learn. Most of the energy we use – in hungry stomachs and gas-guzzling sports cars – disappears expensively and irrevocably (there's more about that in Chapters 13 and 14). Energy is precious, there's only a finite amount packed inside Earth and we should think very carefully before we waste it.

A QUICK WORD ABOUT SPHERICAL COWS

Spherical cows? Let me explain. The Law of Conservation of Energy might be the scientist's equivalent of a well-managed bank account, but the books never balance quite as simply as I've implied. Science can be as maddening as real life. Just as your pay packet is nibbled to nothing by dozens of tiny bills, so the useful energy we can lay our hands on dissipates in a variety of useless ways. In other words, almost everything we do is horribly inefficient: it takes much more energy than we'd expect from simple, scientific theory alone.

When you wolf down a chocolate-chip cookie, the energy it contains isn't instantly absorbed by your body so it can magically reappear as an equal amount of potential energy. As we'll consider in a bit more detail in Chapter 14, much of the energy we eat gets 'lost in translation' and, in practice, only about 20 per cent of it helps us to do what scientists would call useful 'mechanical work' (things like climbing stairs). On the plus side, this means that we can indulge ourselves with chocolate-chip cookies without feeling quite so guilty. On the minus side, as we see in Chapter 5, very similar reasoning explains why cars are so expensive to run – because only about 15 per cent of the energy we squirt into our tanks actually speeds us down the road.

Real life is always more messy than scientific theory – and that holds true for all my other super-simplified examples. If you tried boiling water for coffee by furiously stirring it with a spoon, you'd never manage it, no matter how hard you tried. Heat would escape from the cup as fast as you were adding it with your hand, so there would be no net temperature rise. Hamsters hurtling around wheels wired to dynamos would also doom us to cold coffee; they'd get bored long before the water bubbled and boiled. Vacuum cleaners that use 1,000 watts of power don't give us 1,000 watts of useful suction. Much of the energy they pull in through their power cables disappears in the heat and hum of their motors.

If scientists fretted over every single detail of every problem they have to consider, they'd never get anywhere: simplifying cuts to the chase, offering rough-and-ready answers to things that matter. But it's important not to get too carried away. As our friend Albert Einstein once wisely said, science should be as simple as possible *but no simpler*. Hence the famous joke: what would happen if you rounded up the world's top scientists and asked them a really practical question, such as how to increase milk production in a dairy? They'd scratch their beards for a few days, scribble on their blackboards for a few days more, then proudly reveal their answer, noting (possibly in a tiny footnote on page 97) that it applies only to perfectly spherical cows floating in a vacuum.

▼ *A spherical cow floating in a vacuum.*

CHAPTER THREE
Superheroics

In this chapter, we discover

> *How your thumbs and toes can explain the secret workings of the wheel.*
> *Whether you could burn your house down with an electric drill.*
> *What kitchen knives and golf clubs have in common.*
> *How long a lever you'd need to lift up the Earth.*

You might not feel like a superhero, but that's exactly what you are. You can knock down walls, punch holes through bricks and lift a car with your bare hands. How? Not with your puny human body, of course, but with a bit of help from the scientific tricks packed into household tools such as hammers, jacks and screwdrivers, and powering more complex machines like coffee grinders, washing machines, pneumatic drills and pressure washers.

The human machine

Except for those agonising moments when we snap and shatter bones, most of us would prefer not to think about the skeletons lurking inside our skins; we can do without ghostly reminders of our own mortality. 'Out of sight, out of mind' holds doubly true for the white, inner scaffolding we generally take for granted. Our skeletons aren't just the equivalents of the hidden steel frameworks holding skyscrapers high. They're essential for supporting our own weight, but they also help us to magnify the forces our muscles make so we can walk, run, lift things and generally work with the world around us. Skeletons, in short, are the machines hiding inside us.

In everyday life machines are thing like cranes, bulldozers and engines, or the kind of riveting robotic assembly lines that stitch together jeans and weld cars. In science, however, machines are simpler things – we call them exactly that. A **simple machine** is anything that boosts a force, which includes everything from tiny teaspoons and squeaky wheelbarrows, to ballpoint pens and screwdrivers. Our own bodies work like simple machines because our skeletons are little more than string puppets without the string. Almost every bone and joint in your body works like a lever – your fingers, arms, feet, legs and all the rest. Physical life is all about leverage and, brains and gory bits aside, people are nothing but simple machines.

Nip to your local hardware store and you'll find hundreds of different tools, but almost all of them fall into three or four different types, based on levers, wheels and wedges. These are the main kinds of simple machine. Wheelbarrows, for example, are a type of lever, but so are wheels themselves, and gears and pulleys are levers too. Kitchen knives and chisels work in the same way as ramps you drive up and down on, and so do screws. All the time-saving tricks that tools help us with can ultimately be explained by a handful of scientific ideas.

Why believe in levers?

Archimedes (one of those balding, bearded ancient Greeks who helpfully founded modern science) famously said that if you gave him a lever long enough he could lift the Earth. (According to my calculations, the lever would have been 80 million trillion kilometres or 50 million trillion miles long – *500 billion times longer* than the vast stretch of nothingness between Earth and the Sun[33].) The lever is the father of all machines because most tools are based on levers of one kind or another. A lever is just a rod that pivots near one end, something like a crowbar. The longer it is, the more it multiplies the force you apply; taken to a ridiculous extreme, that's where Archimedes got his planet-lifting logic from. Most of us settle for more mundane kinds of leverage every time we open a door or screw a jam jar tight. Wrenches, spanners and crowbars are obvious examples of levers, but all kinds of other things – including handles, switches and even rolls of paper towel (see box overleaf) – also rely on leverage.

Wheels, interestingly enough, are also levers at heart. You can see that straightaway if you think about taps (faucets) – the handy valves that keep water turned off until you twist them on to release the pressure. Many of us have taps with crossbars that are, of course, tiny little levers: the longer the bars, the easier it is to turn the water on or off because the more leverage you get. If you have arthritis or a disability connected with your hands or arms, you might have special taps with long levers attached that let you turn the water on or off by knocking them with your wrists or elbows. Some of us have more wheel-like taps and, with very little imagination, it's easy to visualise a wheel as just an infinite number of levers arranged in a circle like the petals of a daisy. If your water is operated by a stopcock, that works as both wheel and lever at the same time.

The great thing about levers – and wheels – is that you can use them in two completely opposite ways to beef up either the force or speed you can make with your body. If you turn a stopcock, it's exactly like using a spanner on a rusty nut: you're pushing the free end of a lever around the edge of a circle to produce a slow-moving, extra dollop of turning force in the middle. But you can turn a lever from the other end as well – and that's what happens when you swing an axe. Your shoulders turn at the centre of the circle, while the leverage of the long wooden handle produces extra speed at the edge, walloping the heavy head through the wood. We explore this cunning idea in a bit more detail when we look at bicycles in the next chapter.

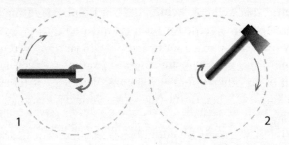

▲ A lever can multiply force or speed – but not both at the same time. *1. When you apply a modest turning force to the end of a spanner (long, thin arrow), the nut in the centre turns more slowly but with more force (short, fat arrow). 2. The opposite happens with an axe. You swing with quite a bit of force at the centre (short, fat arrow) to produce speed at the edge (long, thin arrow). As noted in Chapter 4, gears work in exactly the same way.*

The wheel thing

When it comes to brilliant inventions, the wheel has been on a roll for about 5,000 years. Wheels power us from our homes to other places, but there are plenty of wheels indoors too. Washing machines, egg whisks, electric drills, coffee grinders, computer hard drives, DVD players – all these household things (and many more) rely on wheels. Given how ubiquitous wheels are, and that they're among the greatest inventions of all time, you'd think more of us would understand what makes them tick.

Working like a lever is the simpler of a wheel's two scientific secrets: the bigger you make a wheel, the more leverage it gives you. Spin the rim with a certain amount of force and the hub (the pivot point of the lever) grinds around more slowly but with extra force. That's why, before power steering became popular, old-fashioned trucks and buses had gigantic steering wheels. Wheels have another secret too – and it's much more subtle.

WHY PAPER STREAMS OFF A ROLL

In the UK a long-running TV ad campaign has been helping the Andrex company to sell roughly 13 million km (8 million miles) of 'soft, strong and very very long' toilet paper, every single year, on the back of a charming joke: if a playful puppy pulls on a roll of paper, the whole thing will unravel around your home at dizzying speed.

There's cunning science behind this simple gag. A roll of paper is simply a wheel – and a wheel acts like a lever: the bigger the wheel, the greater the **leverage**. It's much easier to tug paper from a full roll than an empty one because the 'wheel' diameter is wider and you get more leverage at the edge. Once the roll starts rotating, it gains **momentum** – a scientific word that explains why hefty things like trucks and oil tankers take a lot of time to halt. A full roll is more weighty than an empty one, and has its mass concentrated around the edge of the empty tube in the middle (in scientific terms, we say it has a higher **moment of inertia**). Once it's turning, it keeps on spinning all by itself because of its momentum – just like the hefty flywheel on an old-fashioned traction engine. And that's how you end up with paper all over the floor.

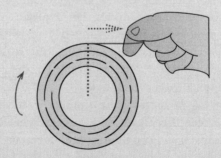

▲ **Spinning a paper roll.** *A roll of paper spins with what's called* **angular momentum** *(solid line) as you pull it. The more paper there is on the roll, the bigger the diameter of the wheel and the greater the leverage (dotted line).*

How wheels reduce friction

Clamp four wheels onto two axles and you have a cart, which is a fine thing for shifting heavy loads and much easier than lifting them by hand or dragging them through the dirt. Everyone knows this, but what's the explanation? How *do* wheels make things easier to move? It's all to do with forces – and how wheels rotate around skinny poles called axles. Imagine yourself lying flat on the floor and being dragged along by a rope attached to a horse. It's going to be fairly painful because the entire surface of your body will be scraping along the ground. The horse is going to find it tough too, because it's working against the force of **friction** (the rubbing between the rough surface of your body and the ground beneath).

Now let's turn you into a human cart. Suppose you extend your two thumbs and your two big toes and point them outwards so we can use them like axles, with four wheels stuck onto them. Assuming you can keep your body tense, rigid and clear of the ground, what's this going to feel like? Instead of your entire body scraping over the ground, you'll just feel a little rubbing as the wheels spin around your thumbs and toes. The huge friction we had before has

▲ **Wheels reduce friction by shifting it to the axles.** *1. Dragged without wheels, the underside of your whole body experiences the unpleasant rub of friction (large hatched area). 2. With wheels slotted on your thumbs and toes, those are the only places that feel a frictional force (four small hatched areas) – and you're much easier to move.*

been cut down to size. We've shifted it from the moving object itself (the bulk of your body) to the axles (your thumbs and toes). That's the hidden secret of how wheels work: they reduce friction by *transferring* it to the axles. It still takes force to move a cart because you still have to work against a little bit of friction, but there's much less to worry about now. And here's where the leverage of the wheels helps. If you're pushing a cart from behind, the wheels act like levers and multiply your pushing force, making it easier for them to grind around the axles and overcome what little friction there is left.

Ramping up

Loading a truck with rubble is back-breaking work if you just put the stuff in sacks and swing it in by hand. It's far easier to pile it in a wheelbarrow and trundle it up a ramp. Wheelbarrows are brilliant examples of machines that we'll come to in a moment, but ramps are machines too. If you imagine how the load is moving, ramps might seem to be working like levers. As you push something up a ramp the load moves up into the air, so you have an end result that's a bit like leverage.

It's easier to understand ramps by thinking about energy. If you have to lift 200 kg (440 lb) of rubble into a truck that's 1 m (3 ft or so) off the ground, you have to use the same amount of energy to get the rubble inside however you do it (the Law of Conservation of Energy tells us that). The smallest amount of energy you can possibly use is 2,000 joules[34]. If the rubble is packed in a sack and you bend your knees to lift it straight up, in about one second, you're using 2,000 watts of power to do it – you're working as hard as a kettle or an electric toaster. However, if the rubble is in a wheelbarrow and you push it up a gentle slope in about four seconds, you can get away with supplying the same 2,000 joules of energy four times slower, at a rate of only

500 watts (that's more like a hand blender). So, forgetting about inconvenient things like friction and the sound energy you're losing in that rusty, squeaky wheel, you're working only a quarter as hard. Because you're pushing the barrow up an incline, you're using less force than if you lift the same amount of rubble straight upwards in a sack. But there's a catch: you have to push the wheelbarrow further (you must use the force over a greater distance) and that means you still have to use the same amount of energy. It's four times easier, but you work four times longer.

Screws are also ramps

If you imagine a pointy hill that's shaped like an upside-down ice-cream cone, with a spiralling road carved into it from the bottom to the top, it might start to look familiar. Shrink your hill to the size of your little finger and make it out of steel, and what you've got is a screw. A screw is simply a ramp that curves around in a spiral – and it works in exactly the same way.

Suppose you want to put up some bookshelves by screwing brackets into a wall. One option would be to pound the screws straight into the plaster, effectively treating them like chunky nails. That would need quite a lot of force – and it would make a horrible mess of your wall. Another option is to do what you're supposed to do with screws: spin them into the wall with twisting turns of a screwdriver. With each turn you're twisting your wrist around quite a distance and the screw moves into the wall by only a fraction as much. Just like driving up a winding road or pushing a wheelbarrow up a ramp, you're reducing the force you need and using less power: it feels easier but it takes you longer. Your hand and wrist are also working like a wheel, in this case, helping to multiply the force you can generate. Some screwdrivers have side-mounted handles to give you even more leverage.

COULD YOU BURN YOUR HOUSE DOWN WITH AN ELECTRIC DRILL?

What happens when you rub one thing against another? Friction gets in the way and generates heat. Use an electric drill for any length of time and you'll whip up quite a bit of friction. Most of the energy that disappears into the drill's power cable ultimately ends up as heat – warming up the wall, the drill bit and the motor itself. It's no coincidence that the earliest (prehistoric) drills weren't used for making holes, but for spawning sparks and setting kindling alight. Most DIY enthusiasts are sensible enough not to touch a drill bit just after they've pulled it from a wall. This immediately prompts the question: could you set fire to your house if you drilled for long enough? Let's do the maths and find out.

How hot can it get?

Suppose you're drilling into solid wood, which usually catches fire at about 200–400°C (400–750°F)[35]. Let's assume the highest possible ignition temperature to start off with. Let's also assume that the wood doesn't conduct heat very well and your drill heats up only a very small part of it where the bit is turning – maybe just 250 g (9 oz) worth. The **specific heat capacity** of wood is about 2 kJ per kilogram per degree, which means that it takes 2,000 joules of energy to heat 1 kg (35 oz) of wood by 1°C. If the room temperature is about 20°C (68°F), we have to heat the wood by 380°C (716°F), so we need to give it about 380 × 2,000 × 0.25 = 190 kJ of energy. If a typical electric drill is rated at 750 watts, it takes in 750 joules of electrical energy each second and (let's assume) puts out 750 joules of mechanical energy, which is all converted into heat. That means we'd need to drill for about 250 seconds – a mere four minutes – to set fire to the wall.

Is it worth worrying about?

DIY suddenly sounds quite risky! But how risky is it? The assumptions I've made might be forgiving. When did you last spend four whole minutes drilling a hole? Not all the drill's energy goes into the wood and the wall immediately next to the drill bit loses heat (passing it on to other bits of the wall), as well as gaining it from the drill itself. I've guessed and rounded off my figures. Then again, what if the drill heated a much smaller amount of wood, or even just the dust in the hole? What if the wood caught fire at only 200°C (400°F)? Then you wouldn't need to drill anything like this long. It seems fair to say that it's *possible* to set fire to a wooden wall by drilling it for long enough. The real risk would be setting fire to the sawdust or splinters you'd produced by drilling, because powders (with tiny particles separated by lots of oxygen) burst into flames much more easily than the materials from which they're made. That's how prehistoric fire-making drills worked, after all – and modern electric drills are far more powerful.

How real tools work

Levers, wheels and ramps are the secret science behind almost every tool you care to think of. Most real-world DIY tools combine two or more of these clever ideas to excellent effect.

Wheelbarrows

A wheelbarrow is a perfect example of several simple machines packed into one handy tool. It pivots around the wheel at the front, so it works like a lever. This means that if you've got a heavy load to move, the best place to put it is right up front near the wheel. The long metal container of the barrow and the handles joined on to it act like a single long lever, making it easier to lift the load into the air. Once you've picked up the barrow and are starting to push it along, you can take advantage of the squeaky wheel and axle at the front. And if you need to empty your load into a truck, a ramp is going to be a big help.

Axes

If you have logs to chop, you'll need to use a hefty axe for the purpose. The long handle is a lever that extends the swing you get with your own arms, pivoted about your shoulders. Your arms, in turn, extend the leverage your upper body gets as it pivots about the base of your spine or, if you swing more energetically, about your feet planted firmly on the ground. Thus there are at least three levers working together and their job is to help the head of the axe to accelerate as quickly as possible, so that it strikes the wood with a great deal of speed and energy – chopping power, in other words. But the head of the axe is working in a different way: razor sharp and wedge shaped, it works like a ramp. As you power an axe into a log, the wood splits apart by moving diagonally along the slope of the blade, instead of just separating horizontally, which means that you can split it in two with much less force. It's exactly the same as pushing a wheelbarrow up an inclined ramp instead of trying to heave it straight up, vertically.

Hammers

Hammers aren't so different from axes: the longer the handle, the more leverage you can get when you swing and the harder you can hit. However, what you're trying to do with a hammer is quite different: you want to pound a nail as far into the wall as possible. A hammer helps you to achieve this in two ways. Because the head of a hammer is bigger than the head of a nail, the force arriving with the hammer blow is concentrated into a much smaller area. Thus the force is magnified enormously and this helps to drive the nail into the wall.

There's another factor at play here. A hammer is much heavier than a nail. Imagine that a hammer and a nail are a ball and a football player's boot, about to make contact. The faster the boot arrives (on the end of a fairly meaty leg, working like a lever to increase the speed of your foot), the more energy it has. We know from the Law of Conservation of Energy that the total energy the leg and boot have before a kick has to be the same as the total energy the leg, boot and ball, combined, have after they collide. If the boot screeches to a halt when the footballer kicks the ball, the ball gets all the energy instead[36]. Because the ball is so much smaller and lighter than the leg and boot, it flies off with greater speed. Exactly the same happens with a hammer and nail: a typical nail weighs 100 times less than a typical hammer, and the vast difference in mass helps to pound it into the wall.

Syringes

Lots of things that might not appear to be tools, in the conventional DIY sense, work in a tool-like way powered by science. Syringes, for example, convert a slow, gentle press of the plunger into a sudden squirt of liquid at the other end. It's easy to see how a syringe works its magic if you know that liquids are virtually impossible to compress. Try squeezing a litre of water into any less space and you'll find that it's practically impossible: that's why a belly flop into a swimming pool hurts much more than landing on a squashy mattress, and why jumping off a high suspension bridge into a river is invariably fatal. The molecules of water resist the pressure in exactly the same way that the hard ground beneath a building prevents it from sinking: smacking into water at speed is much the same as hitting concrete.

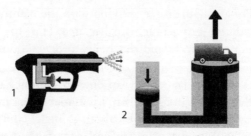

▲ **Water in motion.** *1. With a water pistol (or a syringe), you press on a wide piston with a relatively large force. This makes the water shoot into a narrow chamber and squirt out from the gun much more quickly than you're pressing the trigger, but with less force. The hydraulic jacks in garages work in the opposite way (2). This time you squirt fluid down a narrow pipe. As it enters a wider chamber, it slows down dramatically but produces much more force*[37].

What this means in practice is that water (or a similar liquid) can be used to transmit a force along a pipe. Squirt some water into one end and exactly the same amount will spray out of the other. Make one end of the pipe bigger than the other and you multiply the force accordingly. That's how things like hydraulic cranes and diggers work. A pump powered by the engine shoots liquid down a narrow pipe, which moves a hydraulic ram in or out, forcing the crane's boom (or the digger's bucket) up or down.

SCIENCE AND THE TOOLS OF SPORT

If DIY is not your thing, you might not know one end of a screwdriver from another, or care very much how efficiently a drill works or why. However, you can't escape simple machines that easily. All sorts of kitchen tools (from the simple table knife, which works as both lever and wedge, to egg whisks that use gears and wheels) are based on identical science. Driving a car uses virtually the same principles, from levers, wheels and gears to hydraulic brakes (which are a bit like foot-powered syringes). Even if you're relaxing at the weekend, you're probably putting tool science into action – as you jog or swim, or play football or golf (all of which use arms and legs for leverage).

Leaving aside strategy and tactics, the science of ball sports is all about efficiently transferring energy from body to ball (sometimes using a bat as a go-between), so that the ball goes as far as possible with perfect, positional control. In football you use the leverage of your leg to transfer energy and momentum to the ball. The relative mass of leg and ball determines how effectively the energy is passed over. The longer your foot touches the ball, the bigger the **impulse** – the scientific term for how long we apply a force to something – and the more momentum the ball gets. This is partly why athletes do their best to 'follow through' (keep their limbs moving to extend the moment of contact). Australian sports physicist Rod Cross has calculated that tennis, softball and baseball players get the most effective energy transfer from body to ball when their arms are about six times heavier than the bat and the bat is six times heavier than the ball[38]. That's why a ball shoots off at such speed when you smash it with a cricket bat or a tennis racquet: all the bat's energy has to go somewhere.

One reason why it's so hard to master a sport like golf is that the swing is complex, with numerous different bits of science going on simultaneously. Your entire body is rotating about your hips, giving leverage, at exactly the same time as the club rotates, also giving leverage. The club transfers energy to the ball depending on how well, and how long, it makes effective contact. As in the case of tennis, the relative mass of club and ball is important – and that's before we consider the subtleties of ballistics and aerodynamics

▷

(the angle of flight that makes the ball go furthest, what effect the dimples on the ball have, the effect of backspin and so on). Put all these things together and you have a hefty computational problem that your brain solves mostly by 'practice makes perfect'. Endless 'scientific experiments' help your muscles to learn exactly how much force to apply, when and where.

Science on your team

Sporting champions have no choice but to think in scientific terms: the slightest improvement is a competitive edge. However, even amateurs can benefit. Science works through what we call the **scientific method**, which means coming up with a theory about how the world works, based on things you notice, and gradually refining it using reliable evidence you've gained from experiments. I taught myself to swim by seeing movement through water as a scientific problem. According to Newton's third law of motion, you need to pull water backwards to shoot your body forwards and kick your feet down to stay up. That was all the theory I needed; a few simple splashing experiments showed my theory was a good one. From championship tennis to chopping carrots in the kitchen, the same is true of all sorts of everyday problems. Take a moment to figure out the science – then use it to win.

CHAPTER FOUR

The Beauty of Bikes

In this chapter, we discover

> *How a bike works like a suspension bridge.*
> *Why riding a bike is like kneading bread.*
> *Why cyclists really shave their legs.*
> *What a keen cyclist can learn from a leaping salmon.*

It's easy to cheat at cycling. How about building a bike twice the length of a double-decker bus or three times taller than a man? What about a bike that can carry 24 people at once? Or one that flies in the slipstream of a dragster, faster than an InterCity train? According to the *Guinness Book of World Records*, all these things have been done.

There are much more controversial ways to cheat at cycling. Just ask seven-times Tour de France winner Lance Armstrong, who used performance-enhancing drugs to boost his bike power as he puffed and pumped to victory in the famous yellow jersey[39]. Armstrong's admission of guilt prompted howls of outrage in the cycling community – and rightly so. But what all those shocked cyclists failed to acknowledge is that cycling is *one big cheat*: everything about bicycles is geared towards going faster and further, and travelling more efficiently than you can go on foot. I'm joking, of course, but there's a serious point too. Putting science to good use can help us beat the trials of life, and bicycles are one of the best examples of this you'll find.

Bikes are wheels
Long before you pump the pedals, with your feet barely lifted off the floor, bicycles have started to work their

wonder. Remember how we calculated the pressure on the ground from a typical house and found it not so different from the pressure on our own ankles? How do things compare when you're straddling a bike?

Once your feet are off the ground, your entire weight is balancing on a couple of more or less empty circles (better known as wheels) with a few flimsy bits of metal (spokes) stretched between them. The closer you look at a bike wheel, the more impressed you should be. The first wheels, on ancient push carts, were solid discs of heavy wood, and it's easy to see how they could support any amount of weight. Wood is almost impossible to squash, so the load presses down on the cart and the atoms in the wheel press back up, resisting compression and safely taking the strain. Solid wheels can carry a ton (or several), but their drawback is that they weigh a ton too, so they take an awful lot of effort to push, especially over bumps or uphill. That's largely why spoked wheels were invented: much of the wood was whittled away, leaving a few solid struts to share the load. This time-tested design is still used in car wheels today, though not in bikes.

Empty promise
With bike wheels, there's something different going on. A bike's diamond-shaped frame rests on the two wheel hubs; the hubs have axles running through, with the wheels spinning around them. But take a closer look at those wheels.

Look at the spokes. Each one is amazingly flimsy − not much different from a wire coat hanger that you could easily bend with one hand. It's no exaggeration to say that more than 99 per cent of a bicycle wheel is empty space. That's what you're resting on when you cycle: more or less nothing. Of course the secret is the number of spokes. A typical racing bike wheel has around 24 spokes (some have 32, 36 or 40), making a total of 48 spokes on both wheels. Let's say you weigh 75 kg (165 lb) and the bike itself weighs another 25 kg (55 lb), making 100 kg (220 lb) in total that the spokes have to support. For the sake of easy maths, let's assume that the two wheels can muster 50 spokes between them, so that gives us about 2 kg (4.4 lb) resting on each spoke. Could you really balance two bags of sugar, which is roughly what 2 kg looks like, on top of a single wire coat hanger without any risk of it bending? It sounds most unlikely to me. In scientific terms, therefore, there must be something else going on as well.

Bike or bridge?
You can easily bend a bicycle spoke, but you can't *stretch* it no matter how hard you try. That's the key to how a bike wheel works. Unlike a conventional cart wheel, which is compressed, the spokes of a bike wheel are taut (we say they're in **tension**) like the strings of a violin, the threads of a spider's web or, perhaps the best comparison of all, the cables in a suspension bridge. If you consider that the bike's entire weight presses down on the hubs, the way most of the weight of a suspension bridge presses down on the deck,

▲ **Bike bridge.** *Take one suspension bridge and lose the two towers (light grey). Shrink the deck (dark grey) to a small circle in the middle. Tie the drooping steel hawsers (black) together at the ends. Then make the thinner suspension cables neat and evenly spaced. What you'll end up with will be something like a giant bicycle wheel. Both of these ingenious structures work through tension.*

you can see that a bike effectively hangs from its spokes in the same way that a bridge deck hangs from the cables overhead. At any moment the spokes above the hub are slightly more stretched than the ones below, but all the spokes are in tension, all the time.

Even more interesting than the flimsy design of the spokes is the way they're scattered around the hub. Unlike in a simple cart wheel, the spokes in most bike wheels don't run directly from the rim to the centre of the hub. Instead, each spoke stretches a little bit further to the side of the hub, making what's called a tangential connection. Because the wheel hub is fairly wide, some of the spokes connect to one side and some to the other. What we get, then, is a kind of taut mesh of spokes with the bicycle's entire load spread evenly across it. It can resist the weight of the bike and its rider, but also (in the case of the front wheel) the shearing, twisting forces as you steer to the left or right at high speed or lean into a bend. It's a powerfully tensioned, three-dimensional structure that can resist forces in all directions. When you consider that each basic component of the wheel – each flimsy little spoke – is so weak that you could bend it in one hand, it's clear that bike wheels are nothing less than engineering miracles.

Bikes are levers

Spokes and wheels are just the beginning. Want to see some more cheating? Remember the 'unfair' advantage that leverage gives us when we want to magnify the forces our body can produce – then count the number of levers you can see on a bike.

The handlebars are levers, for starters, making it easier to twist the front wheel even when the tyres grip the road and you're haring along at top speed. It's far easier to steer a mountain bike, with its long, straight handlebars, than a racing bike, with its shorter, curly handlebars designed to force your elbows into a more aerodynamic posture. That leverage has another handy benefit. Wider handlebars make it easier to keep the front wheel pointing forwards when you meet bumps in the road. Suppose you hit a pothole that jerks the front wheel suddenly to one side. The force you feel at your arms is considerably reduced by the leverage of the handlebars, so the bike is easier to hold in a straight line.

The pedals (more specifically, the cranks – those straight-line rods going up and down from the flat bits that hold your feet) are also levers, which help you take advantage of the leverage in your own legs as they pump up and down. Clearly both handlebars and pedals have to compromise on leverage to a certain extent. In theory, the longer the handlebars, the easier it is to turn your bike (bigger handlebars work like a bigger steering wheel on a bus or a truck). And the longer the pedals, the easier it is to kick them around. However, the handlebars can only ever be as wide as your outstretched arms, while the pedals obviously can't be so long that they bump the ground as they spin around.

The need for speed
The wheels of a bicycle are also levers, although you might not recognise them as such straightaway. As we saw in the last chapter, most levers – cunningly disguised as tools such as hammers, crowbars and wrenches – are designed to increase force: you turn the outer edge a longer distance with as much force as you can and the centre turns a shorter distance, more slowly and with correspondingly more force. In short, an ordinary lever gives you more force but less speed. Generally, we hop on bicycles because we want to go somewhere faster than we can by walking. In bicycles, we want the wheels to work in the opposite way to traditional levers – and that's exactly what they do.

If you hold your arm straight out in front of you and turn it through 90 degrees, as though you're bowling a cricket ball, overarm, your hand sweeps through a much bigger arc than your shoulder. It goes further in the same time so it goes faster and (although it might not be apparent) with less force.

That's why taller tennis players with longer arms generally hit the ball faster than shorter, stockier ones. In exactly the same way, powering the centre of a wheel gives you more speed at the edge – and that's how bicycle wheels work. The bigger the wheel, the more the leverage, and the greater the speed you get, although at the expense of the force you produce to move you along. That's why racing bikes, which need speed, have bigger wheels than mountain bikes, which need less speed but plenty of force for climbing hills and bucking bumps. Penny farthings, among the earliest bikes, had gigantic wheels to multiply your speed accordingly, but they weren't much easier to ride than elephants (who wants to climb a ladder to get on a bike, and what if you fall off?).

Bikes are gears

If you live where I do, 100 m (330 ft) above sea level and 10 minutes from the coast, nipping into town and back by bicycle is a lot less appealing than walking or taking a bus. It's downhill all the way – and uphill all the way back. If the

▼ **Wheel leverage.** *Wheels are levers turning in a circle that continually multiply either speed or force. The speed magnification is easy to understand. If you turn a wheel through 90 degrees, at the centre, you can see – by comparing the length of the two arrows – that the edge must go faster simply to keep up. It's less obvious that it can multiply force, but if you imagine this is a stopcock you can probably almost feel how much more bite there'd be at the centre if you cranked the rim.*

world were completely flat, using a bike for short journeys like this would be an obvious choice, but those pesky hills do have a habit of getting in the way.

Bicycles such as the penny farthing were designed entirely for speed: there was no way you could pedal them uphill even if you wanted to. Traditional bikes like this use leverage in only one direction. You pedal-power the centre of the front wheel and the rim turns faster but less forcefully, draining the 'bite' you need to climb hills. It's easy to see how to make a fast bike – just add big wheels – but what we really need for going uphill is the opposite: a bike with ultra tiny wheels that somehow multiplies our pedalling force. Pedal the bike quickly and the wheels turn slowly, but with enough force to climb. It sounds perfect. So, living where I do, what I really need is two bikes: one for going downhill into town that multiplies my speed and reduces my pedalling force, and a second one for coming back up the hill that reduces my speed and multiplies my force for climbing. Could we somehow pack both machines into a single bike? Yes – and that's exactly what gears do.

How do gears work?
Gears are simply wheels of different size with teeth around the edge so they can 'mesh' – lock and turn in step. A pair of gear wheels will either increase the speed of a machine and reduce the force it can produce, or do exactly the opposite; it can boost speed *or* force but not both at the same time.

At first sight you might think gears were all about teeth, but the nibbled outer edges of the wheels merely stop them from slipping and are otherwise irrelevant. The secret of a gear lies in the relative size of the two wheels that mesh. If one wheel works like a lever, two wheels touching one another at the (shared) rim work like two levers touching. If you turn the centre of the first wheel, its rim turns with more speed and less force. The second wheel, which is touching the first wheel, must turn at the rim with exactly

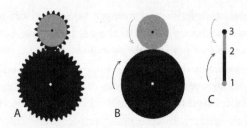

▲ **How gears work.** *Two gears that mesh together (A) are just like two wheels of the same size that touch at the rim (B) – and two touching wheels work like an infinite number of touching levers (C). If you rotate the black lever around the axle at point 1 with a certain speed and force, due to the length of the lever point 2 will be turning faster but with less force. When the black lever touches the grey one, the leverage works in reverse. As the black lever pushes the grey one around, the axle at point 3 rotates as well, more slowly than at 2 but with more force. Overall, the two levers result in the axle at 3 turning faster but with less force than the one at 1. Exactly the same thing happens with the gears on the left.*

the same speed as the first wheel. The centre of the second wheel then turns more slowly and with more force than the rim. If the first wheel is bigger than the second wheel, what we get is an overall increase in speed and a reduction in turning force (which is called **torque**). If the second wheel is the bigger of the two, the first wheel turns faster and the second wheel goes slower, but with more torque.

On a bicycle, the back wheel is a fixed size and so is the gear to which the pedals and their cranks are directly attached. In theory, then, there's a fixed relationship between the two wheels, which means that a bicycle could only ever help you in one way: by choosing gear wheels of a particular size, you could set it up either to increase speed as you pedalled or to increase force, but not both. In other words, you could build it to race along the straight at super speed or climb hills at a grinding pace, but not both. We're effectively back where we started – needing two different bicycles to go uphill and downhill.

Fortunately, bikes have another trick: the gears on the back wheel and the pedal wheel are connected by a flexible chain, and there's a whole selection of gear wheels of different sizes on both wheels to choose from. By flicking the gear lever, you can flip the chain onto different pairs of gear wheels so, effectively, you're changing the relative size of the two gear wheels connected together. A cunning mechanical gear-shift called a dérailleur allows you to do this as you're bicycling along, even while the gear wheels and the chain are spinning at speed. Changing gears makes it possible for one bicycle to work in two utterly opposite ways. In a high gear, for racing down the straight, the back wheel will turn more quickly than the wheel you pedal (faster and with less force). In low gear it's the opposite, with the back wheel turning more slowly but with a huge dollop of extra force for digging in and crawling up hills.

How do the numbers work in practice? If you're an Olympic champ belting along at speed, your gear ratio (the number of teeth on the back wheel divided by the number on the pedal wheel, or chainwheel) could be as much as 1:5 – so, effectively, you're multiplying the speed of your back wheel by five. In other words, if your wheels are about 62 cm (24½ in) in diameter, and they have a circumference of almost 2 m (6½ ft), every turn of the pedals is whizzing you about 10 m (35 ft) down the road[40].

Look, no cheating!

Now just like DIY tools, the various 'cheats' built into a bike – the levers, wheels and gears – have to obey the basic laws of physics. Bike gears can give you extra speed or extra force, but not both at the same time. If they could do both, you could get more energy out of a bike's back wheel than you put in at the pedals and, as we discovered in previous chapters, that simply isn't possible: it violates the Law of Conservation of Energy.

It's easy to see how we're staying within the rules if we compare force, speed and energy at the pedals and at the back wheel. Suppose you're going along the straight in a high gear. Maybe you kick the pedals around once (say) and the back wheel turns around twice, so you're getting a two-fold increase in speed and a halving of force. At the pedals, you're pushing down with a certain force and, each time the wheels rotate, using a certain amount of energy to do it. Now the energy you need to do something (like lifting an orange) is equal to the force you use (the weight of the orange) multiplied by the distance over which you use that force (how high you lift, in the case of the orange). At the wheels, the force is halved but the speed is doubled. A doubling of speed means you're going twice as far in the same time. So you're using half the force over twice the distance, which is the same as using your original pedalling force over the original pedalling distance – and involves exactly the same amount of energy. Hurray, the laws of physics survive again.

Why is cycling such hard work?

One of the great things about cycling is that it's incredibly efficient. Where a car weighs maybe 20 times as much as you do, a bike is relatively light compared to your own body (perhaps just a fifth to a quarter of your weight). In other words, a typical car weighs about 100 times more than a typical bike even though, if you're the only occupant, it's carrying exactly the same load[41]. Riding a bike up a hill merely involves heaving a few tubes of aluminium alloy, a couple of rubber circles, a bit of plastic and a few dozen spokes a certain distance into the air, whereas driving a car up the same hill means shifting 1 or 2 tonnes of steel against the force of gravity. Try pushing a stalled car up a steep hill (as I once did) and you'll quickly discover the difference.

Cycling sounds like a breeze compared to motoring, and it certainly can be. But if you watch professional cyclists or mountain bikers hurtling along for all they're worth, it's quite clearly hard work as well. Is it obvious why? 'Hard work' means that, despite the light weight of their bikes, cyclists are having to expend considerable amounts of energy. Where does it all go? You might suppose we lose energy when we cycle up hills, but that's not really true. If you cycle up a hill, you work against the force of gravity but you gain potential energy in the process. That means you can whiz back down the other side with no effort at all, turning your potential energy back to movement (kinetic energy) and losing relatively little energy. Yet cyclists and cycling clearly do use energy irrevocably so, again, where does it all go? There are actually three main ways you use – and lose – energy when you cycle: rubbing, whooshing and kneading, technically known as friction, drag and rolling resistance.

Friction

As we discovered in the last chapter, wheels work by shifting friction from the road to the axle. As you ride along you're wasting energy pumping the pedals around their axles, making the gears spin on their axles and turning the bike wheels around their axles; a certain amount of your energy is disappearing to friction in all those places. When you pull on the brakes you're pressing lumps of hard rubber against the insides of the wheels to slow them down. Your kinetic energy turns to heat, warming up the rubber brake blocks and the wheels themselves – and that heat is more lost energy. All this energy that you lose to friction is lost with a capital L: you can't use the heat in the brakes and the wheels to do anything useful.

Drag

One of the great joys of cycling is feeling the wind against your face, but that's another way in which you're losing energy. When you're walking, air might seem like a load of invisible, empty nothingness that serves no purpose other than to let you breathe. But it's no vacuum: it's full of molecules getting in your way. Walking through water is really hard because you have to force your body through thick, gloopy liquid. Cycling through air is exactly the same and the difference is just a matter of degree: you're not working quite as hard as you would in the water or wasting as much energy, but you're still wasting *some*. The faster you go, the greater the **air resistance** (or **drag**) and the more energy you waste. On a racing bicycle, at high speed, roughly 80 per cent of the energy you pump into the pedals is used to 'whoosh' your way through the air. On a mountain bike, because you're going much more slowly and bumping up and down on a rough track, only about 20 per cent of your energy disappears this way[42].

Rolling resistance
So where does the other 80 per cent of the energy (in the case of mountain biking) or 20 per cent of it (on a racing bike) actually go? Here's a clue: have you ever tried kneading bread? You have to pick up the dough and endlessly fold it over on itself, squashing and re-squashing it for several minutes. It's surprisingly hard work because you're having to rearrange lots of molecules inside the mixture, pushing some together, pulling some apart and leaving the dough in a very different form from how it was when you started. It might look the same, but internally it's very different: you've done lots of work on it and you've used lots of energy. Riding a bike is similar, because each time you make the wheels go round, the tyres and the air inside them have to be stretched (at the top) and squashed (at the bottom) in a similar way. It takes energy to make tyres turn, which is why we say they have **rolling resistance**. Where kneading uses energy by making dough increasingly springy, riding a bike doesn't transform the tyre in any way, but simply converts energy to heat (and a little bit of sound) by stretching and relaxing it over and over again. Generally, thick and chunky mountain bike tyres have greater rolling resistance than thin, svelte racing bike ones. And that's where the rest of your energy goes. On a mountain bike you'll lose 80 per cent of your

energy to rolling resistance; on a racing bike 20 per cent of your energy vanishes that way.

Energy crisis?

Is there anything we can do to waste less energy on a bike? With three different ways of losing energy, it follows that there must be at least three different ways of saving it.

Fighting friction

At first sight friction seems to be the easiest of the three problems to tackle: you can simply oil your gears and chain. However, these frictional losses – the basic rubbing of wheel against wheel, gear against gear – are the least of your worries. There are much bigger frictional losses when you brake and squander all the momentum you've built up by converting it irrevocably to heat. Experienced cyclists try to get around that problem by minimising the need for braking. You can usually anticipate when you'll need to come to a halt (those traffic lights at the end of the road) and stop pedalling in advance. However, while that certainly reduces your braking, it doesn't prevent you from losing your energy entirely when you come to a halt. There's really nothing you can do about that. Hybrid electric cars and electric trains use regenerative braking, which captures the energy you'd waste putting on the brakes and sends it back to a battery for reuse. As we'll see in the next chapter, that's extremely efficient for big and heavy vehicles that move at high speeds (the energy you can capture and reuse is very significant), but it doesn't really work for light vehicles moving at low speeds with low energy (bikes, in other words).

Smoothing your way

You've probably seen keen amateur cyclists taking to the weekend streets, squeezed into tight Lycra and puffing away under helmets shaped like teardrops. These people are doing their best to cut the 80 per cent of their energy that's being lost in the war against air resistance. Notice how they cram themselves into a cosy peloton (the tight pack up front in a race)? Sneaking tight inside another rider's slipstream saves around a quarter to a third of the energy you'd need if you rode all by yourself[43]. And that's just the start. Professional racing bikes have all kinds of aerodynamic touches, like a second set of handlebars you can operate with your elbows tucked into your body, and bladed spokes on the wheels to cut drag as you steer and lean. Bending your body like a tortoise makes you zoom along like a hare, and all that skin-tight clothing can shave seconds off your time, although it makes rather less difference to the average Sunday cyclist than to the far from average Olympic champ.

Speaking of shaving, what about the idea that taking a razor to your hairy legs can help you go faster? Despite a great deal of searching through scientific journals, I've found no studies proving that shaving your legs has any effect whatsoever – and that's hardly surprising. How would you do such a study? You can hardly cycle through a race, shave your legs, then repeat your performance. And you can't really shave only one leg to compare it with the other. It's just possible to imagine some sort of cunningly contrived experiment in a wind tunnel in which a tailor's dummy, with clippings of hair glued to its legs, has its air resistance measured and compared with a similar dummy left smooth. Perhaps there's a psychological boost from shaving legs, and doubtless there are benefits to being clean-shaven if you fancy a post-race massage or if you tumble from your bike, cut yourself and need medical attention. For everyone but Olympic cyclists riding time trials, however, the air-resistance benefits are surely minimal[44].

Thinking like a fish
Salmon can leap through water because they're long, sleek tubes that sit parallel to the flow; cyclists go fastest when they try the same trick. The best way of cutting air resistance is to swap your bone-shaking upright bike for a lazy,

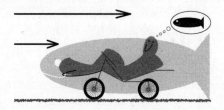

▲ **How recumbent bikes work.** *Recumbent cyclists operate a bit like fish. The prone, tube-like body position dramatically reduces air resistance. Compared to a conventional racing bike, a recumbent uses about 15 per cent less power overcoming drag at speed[45].*

relaxing recumbent – one of those weird, ground-level bikes you ride lying backwards, like a kind of hammock on wheels. Recumbent bikes are the fastest of them all. Because you're cycling in a tube-like position, you slide through the air like a salmon leaping up a river rather than smashing through it bolt upright with all the grace of a removal truck.

If you take your cycling seriously it's easy to get obsessed with such things. The leg-shaving, Lycra-clad blurs that sweep down your street are evidence enough for this. However, it's important to remember the bigger picture. Bicycles are cunning tricks that sweep us from place to place with almost unbelievable efficiency. Using science as their cheat, they're streets ahead of electric cars, motorbikes, diesel cars, steam engines – and even the ordinary human body. Cycling is much more efficient than walking because you're not constantly adjusting your gait, as you have to when you walk, so you waste less muscle energy moving forwards over the same distance. Indeed, you lose so little energy on a bike that you're already well ahead of the game; to me, it seems churlish to quibble about squeaky brakes, lumpy tyres – or even hairy legs.

CHAPTER FIVE
Car Crazy

In this chapter, we discover

Why our ecofriendly world still loves dirty old petrol.
How far you can drive on a teaspoon of fuel.
Why a car uses 250 times more air than a bicycle.
How an alligator's worth of force prevents you from skidding when you round a bend.

One billion cars on the planet[46]. Count them! Stack them on top of one another and you'd get a pile 170,000 times higher than Mount Everest – tall enough to reach to the moon four times over. Park them bumper to bumper and they'd stretch across the United States more than 1,200 times. A billion (a thousand million) is the kind of number most of us find hard to picture. To put it in some kind of perspective, the world's population is a smidgen over 7 billion; between us, we guzzle 2 billion cups of coffee a day. There are roughly 6 billion mobile phones in circulation[47] – and anywhere between 1 and 2 billion sheep[48]. If you can get past a mental image of a sheep tucked behind the wheel of a Jaguar, bleating into the mobile phone that's clamped, cloven hoof-style, to its jaw, can you possibly imagine a billion cars?

The interesting thing for me isn't that there are so many cars in the world, but *why*. What is it about cars that's made them, in the space of just over a century, one of the most successful inventions of all time? Not surprisingly, the answer is all to do with science.

What's so good about cars?

Cars are chemistry labs on wheels. And though that might not sound so interesting, it accounts absolutely for their ubiquity. Take away the leather seats, the gleaming chrome, the go-faster stripes and all the rest, and you're left with a handful of tin cans called cylinders, where petrol explodes into power. Cars are built around engines, and engines (or internal combustion engines, to give them their full name) burn petrol with oxygen in the air to release the energy locked inside it. We think of burning as a way of making fire, but it's essentially a chemical reaction between oxygen and fuel that just *happens* to produce heat and fire as a by-product. The simple science of cars is so utterly mundane that we scarcely give it a thought: pump the petrol in your tank, turn the key and off you go. Think about it more closely, however, and you'll see how astonishing it really is.

Suppose a typical modern family car does about 40 miles to the gallon or, in metric terms, 100 km for every 7 litres of fuel. That means if you have a teaspoon of petrol (about 0.004 litres), it contains enough energy to roll your car about 60 m (200 ft), or roughly 15 times the car's own length. Consider how hard it is to push a car, even once you've got it going from a standstill, and I'm sure you'll agree that's quite remarkable. The simple fact is that petrol is absolutely chock full of energy: short of uranium (nuclear fuel), it's just about the most energy-rich material there is. That, more than any other reason – including the freedom, independence and social status that cars give us – accounts for their popularity.

DEEP BREATHS

Motorists learn never to go far without glancing at the fuel gauge: the Law of Conservation of Energy tells us cars go nowhere without the energy packed in petrol to power

▷

them along. What's much less obvious is that cars need air to breathe, just like people. The combustion that burns fuel in the cylinders is a chemical reaction between the **hydrocarbons** (molecules built from carbon and hydrogen) in petrol and oxygen in the air, so even if you've got a full tank of petrol, if there's no air around you're going nowhere fast. How much air does a car need, exactly? A sports car pants its way through something like 6,000 litres of air (6 cubic metres/1,300 gallons) per minute, which is about 250 times more air than a cyclist uses[49]. So if you drove your car continuously for about eight hours, it would breathe enough air to fill an Olympic-sized swimming pool.

You might think air is just an airy-fairy consideration because, wherever you go on Earth, there's always plenty to spare. In theory, the only engines that ever run short of air are the ones attached to space rockets. Since they blast out of Earth's vast atmosphere, into the deep darkness where there's no oxygen, they have to carry their own 'air supplies' (oxidisers) in giant tanks, as well as their own fuel.

Is there anywhere on Earth where a car could run out of air before it ran out of petrol? It's not very likely, but high altitude – and 'thin' air (low oxygen) – certainly makes a difference to how well a car runs. That's simple science for you: if there's a chemical reaction between A (fuel) and B (oxygen) to produce C (energy), and there's less B, you're going to get less C too unless you compensate somehow. Indeed, quite a few manufacturers have offered modified versions of their cars for mountain-top driving[50].

Generally, our bodies don't much care for high altitudes either. If you're pounding through the clouds in a mountain-top marathon, there's less oxygen nipping down your nose to power your legs: respiration needs that all-important gas just as much as the cylinders in your car do. Long-distance runners have a positive disadvantage at high altitudes because they breathe in so much oxygen during a race.

Interestingly, however, the same doesn't apply to sprinters. Because their races are shorter, they don't breathe for so long, and because the air is thinner, so reducing aerodynamic drag, they can actually run faster at higher altitudes. This is why a number of world athletics records were set during the 1968 Mexico City Olympics (at an altitude of about 2,250 m/7,400 ft)[51].

What's so bad about cars?

Driving a car is the next best thing to being a human cannonball. It can shoot you over the ground at a blistering speed for an amazing distance on a single tank of fuel. If your tank holds 70 litres (about 15 gallons), and your engine can manage 100 km (62 miles) on 7 litres (1½ gallons) of fuel, a single fill-up at the petrol station will power you a rather surprising 1,000 km (600 miles). So, stopping to refuel four times, you could just about drive across the United States from New York City to Los Angeles.

That might sound impressive, but it's nowhere near as good as it might be (or should be). If you want to get across the States, a car is probably a much better bet than the bicycles we considered in the last chapter – at least if you want to expend minimal effort and make the trip as quickly as possible without flying or taking the train.

However, suppose you wanted to climb Mount Everest and such a thing were possible on foot, by bike and by car. Instantly, we find ourselves wondering 'Do I really need to drag all that metal to the top?' Carrying a bike would be bad enough, but if you drive a car up a hill, you're lifting not just your own weight (something like 75 kg/165 lb), but the weight of the vehicle as well (which could easily be 1,500 kg/3,300 lb). This is the real drawback of driving a car. Wherever you go, it's like having a ball and chain shackled to your leg, except that the ball is about 4.5 m (15 ft) long and weighs 20 times as much as you do. If you chug to the top of a mountain, that means about 95 per cent of the energy you need is wasted lifting the sheer bulk of the car. Only 5 per cent does the useful job that you actually care about: moving your own body to the summit. That's why cars gobble down so much more fuel and air than cyclists. What applies to climbing mountains applies equally well to any other kind of driving: you're still shifting extra metal and wasting energy, wherever you go. It's no coincidence that the Ariel Atom, one of the world's fastest cars, is also one of the lightest. It weighs less than 500 kg (1,100 lb), which is about a quarter to a third the weight of a typical small car[52].

In short, there's a basic inefficiency to petrol-powered cars (all that extra weight) that we simply can't dodge. And that's before we go anywhere near considering their real inefficiency. The fundamental problem with cars is that a mere 15 per cent of the energy locked in petrol actually moves you down the road. The rest is wasted in various ways, including heat

▼ **What makes cars so heavy?** *Three-quarters of a car's weight comprises steel, iron and aluminium. The steel body accounts for up to a third of the total weight, while the iron engine adds another 15 per cent or so*[53].

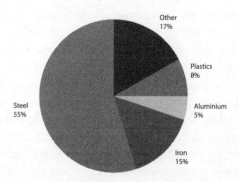

Other 17%

Plastics 8%

Steel 55%

Aluminium 5%

Iron 15%

losses in the cylinders, frictional rubbing in the gears, the sound the engine makes, powering the electrical system and much more besides. If cars were 100 per cent efficient, and all the energy in the petrol were perfectly converted into kinetic energy blasting you down the road, you'd be able to go five to ten times further at least – well over half a kilometre or even more on every teaspoon of fuel.

The more people you pack into your car, the bigger the useful load you're carrying compared to the useless weight of the vehicle, so the greater the efficiency you're getting. That's why things like trucks, buses and trains work out as efficient forms of transportation even though they're bulky and use heavy diesel engines. However, no matter how efficient you make a car, or any other vehicle powered by internal combustion, it's still burning petrol and belching

▼ **Where does a car waste energy?** *Cars are hugely inefficient. Driving in the city, only about 15 per cent of the energy in the fuel you buy (black slice) produces useful power at the wheels. All the rest is wasted in engine losses (such as heat loss in the radiator), 'parasitic' losses (caused by things like the alternator, which steals energy to produce electricity) and losses in the drivetrain that carries power to the wheels. Based on figures from US Department of Energy Office of Transportation and Air Quality.*[54]

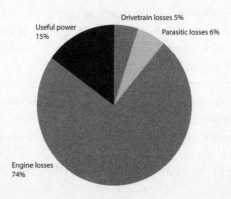

Drivetrain losses 5%

Useful power 15%

Parasitic losses 6%

Engine losses 74%

out pollution of one kind or another, from soot or smog to the carbon dioxide implicated in global warming. How, then, can we build a better, cleaner, more efficient kind of car? What kind of pointers can we get from science?

Better than petrol?

Wacky inventors have come up with all kinds of ways of powering cars. Before petrol came along there were steam cars, for example, but steam engines are by far the most inefficient types of power – and coal is heavy, filthy and spews out smoke. Diesel engines, which are like industrial-strength petrol engines, are almost as old and work on the same basic principle. Although we tend to think that electric cars are radically modern, they date back to well before Henry Ford's time in the late 19th century. Ferdinand Porsche, best known as the father of the modern luxury sports car, originally made his name pioneering hybrid electric cars in 1900[55].

Although it's easy enough to sketch a sleek new kind of car on paper, it's much harder to come up with a credible design that moves you as quickly or as far as petrol. You might think 100 years or more of progress would have made electric cars supremely better than petrol ones, but there's a basic problem: batteries don't pack in anything like as much energy as liquid fuels such as petrol, kerosene (aeroplane fuel) or alcohol (used to power rockets). Even a lump of wood or a bag of sugar holds more energy than the equivalent weight of rechargeable batteries. Moreover, while you can completely refuel your petrol-powered car in just a minute or two, fully recharging the bank of batteries in a silent, electric runaround will keep it docked and useless for several hours at a time[56].

Environmentalists love to imagine that a giant, petrol-headed conspiracy has kept electric cars whistling and waiting in the margins of technology, while dirty, expensive

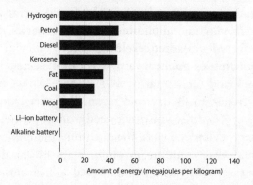

▲ **Why we don't all drive electric cars (yet).** *Kilo for kilo or pound for pound, batteries can carry only a fraction as much energy as hydrocarbon fuels such as petrol and diesel. Hydrogen (top) is the best fuel-carrying substance by far, but it's a flammable and leaky gas, so it's hard to store and transport safely and efficiently[57].*

gas guzzlers continue to pollute the planet. The truth is more prosaic and less sensational: petrol is – and for the time being will continue to be – a far more effective and far more efficient medium for carrying energy than batteries. Science, not politics, is the simple reason why most of us are still driving petrol-powered cars today.

Our electric future?
We won't necessarily be driving petrol-powered cars tomorrow. No one can predict with any certainty when oil will run out – by which we mean when it will become so expensive that market forces make the alternatives more attractive. That day will, however, arrive eventually: it's taken hundreds of millions of years to make our planet's entire oil supply from rotted plants and sea animals, but little more than a century to use up virtually all the oil at our disposal. Oil is being formed every day; assuming we stopped using it tomorrow, if you came back in millions more years, you'd find plenty of new oil underground to drill to the surface

and burn. So, regardless of the fact that batteries aren't as good at carrying energy in portable form as petrol, that's the way the future is heading, like it or not.

Bulky batteries might be a drawback of electric cars, but these sleek and silent, sparky vehicles have plenty of things in their favour. In theory, they're much lighter than petrol-powered cars because you don't need that monstrous engine, those fiery cylinders with their pistons pumping up and down, and that grinding gearbox. In practice, of course, what you need instead is almost as bad: a load of extremely heavy batteries. Even so, electric cars generally work out lighter, which makes them more efficient.

Recycling energy

One of the things that makes petrol-powered cars so inefficient is all the stop-start driving you have to do in cities. As we saw in Chapter 2, it takes energy to do anything at all. If you've ever pushed a broken-down car, you'll know how back-breaking it can be simply to overcome its **inertia** (the basic laziness of a lump of mass) and get it rolling. If your car weighs a tonne and a half (1,500 kg/3,300 lb) and it's grumbling through the city at 65 km/h (40 mph), it has quite a bit of kinetic energy. Do the maths and you'll find it's about 240 kJ, which (according to the figures we used in Chapter 2) is nearly enough to climb the Empire State Building.

That might not sound so very much, but here's the snag. Every time you stamp on your brakes to dodge children chasing footballs or cats unacquainted with basic road safety, those 240 kJ disappear into thin air. As the brake pads meet the brake discs and bring you shuddering to a stop, all that motion energy vanishes in a squeal of tyres and a puff of smoke. In sports cars and Formula 1 racers, the brakes can sizzle up to temperatures of 750°C (1,400°F) – hot enough to set them on fire if they were made of wood[58]. When you step on the accelerator after braking to a standstill, the engine has to pick up speed all over again by turning more petrol into power. So the horribly wasteful cycle repeats itself, over and over again.

Electric cars have a big advantage here in that they're powered by motors. In its simplest form an electric motor is a fat core of copper wires that whirls around inside a hollowed-out magnet. Feed electricity into the wires and they generate a temporary magnetic field that repels the magnet's own magnetism. The copper core spins around and we can use it to power anything from a vacuum cleaner to a bullet train. The great thing about electric motors is that you can run the whole process in reverse. If you twist the shaft of an electric motor with your fingers, you get electricity zapping out of the copper wires in the opposite direction: the motor, in other words, becomes an electricity generator. In theory, you could make electricity by taking any electric appliance (a vacuum cleaner, say) and operating it by hand so that the motor rotates, which would pump electricity out of the other end. So if you unplug your vacuum cleaner and give it the kiss of life, the wacky theory is that useful electricity should come zapping out of its power cable; of course, in practice, it won't work in a vacuum cleaner – but it *does* work in an electric car.

Electric cars use their motors to great effect. When you're driving along, the batteries pump power through the wires into the motors, making the wheels spin around. When you hit the brakes you cut the current, but the car's momentum keeps the wheels turning. Because the motors are still spinning as well, they start to generate electricity, which feeds back into the batteries and slows the car down. Instead of wasting energy when you brake, an electric car saves at least some of its energy by recharging its batteries. It's called **regenerative braking** and it improves the efficiency of a typical electric car by 10 per cent (electric trains, by comparison, manage about a 15 per cent improvement, which is the equivalent of running one in every seven trains for free)[59].

What would a perfect car look like?

Suppose you set out to design a car that's as efficient as you can possibly make it. What would you end up with? You'd want something with as few moving parts as possible, so that it wasted little of the energy you fed in. Ideally, it would be as light as possible too, so you wasted less energy shifting metal, plastic and glass. It would need to run off a widely available fuel that carried piles of energy per kilogram – probably something carbon based and organic. If you didn't mind low speeds (say a modest 6 km/h or 4 mph) and were prepared to consider fuels like fat or vegetable oil, you'd find that the ideal car looked much like your body. It would be low maintenance, energy efficient and easy to park. It wouldn't rust or lose value, and – most of the time – would age rather beautifully.

SLIP OR GRIP?

Every time I see cars screeching around bends, tyres squealing like teenagers on a roller coaster, it amazes me that they don't lose control and spin off sideways. Keeping a car going in a straight line is easy enough: with properly balanced wheels, it virtually drives itself. When you race around a corner, science changes everything. We like to think there's an evil conspiracy to pull us off the road – something called **centrifugal force** tugging us outwards to the far side of a bend. In fact, a car nipping along at high speed has a tendency to keep going in a straight line. Newton's first law says that things keep moving with a steady speed, in the same direction, unless a force acts on them. As you turn the steering wheel, you supply an inwards–pushing force called **centripetal force** that tugs the car around the curve.

The centripetal force comes from friction between the car's tyres and the road. Now you might think tyres are big, heavy things ribbed with hefty grip, but the amount of rubber touching the road from each tyre, at any given moment, isn't that much bigger than the grip you get from one of your own shoes. Think about that the next time you screech home to be in time to put your children to bed: there's barely four shoes' worth of rubber between your life and death[60].

What sort of forces keep a car on the road? There's a simple equation for working out the force you need to make something go round in a circle. If your car weighs about a tonne and a half, and you're driving at 100 km/h (60 mph) around a reasonably shallow bend, you're looking at a force of maybe 10,000 newtons. Going back to our table in Chapter 1, that's about the same force that you get from an alligator bite. So it's a chomping jaw's worth of force that clamps you to the road.

Well heeled

Why don't tyres wear out more often? Because with careful driving, under normal circumstances, they never actually slip on the tarmac. As you drive along, the wheels turn around their axles: that's where the friction happens. The tyres grip the road: there's little or no rubbing there. Car tyres work exactly the same as tank tracks. When a tank whizzes along, the tracks are laid down in front of the wheels, then picked up again behind them. A car tyre is exactly the same, except that it's wrapped all the way around the wheel. The rubber is laid down just ahead of the wheel and picked up just behind it. Unless you brake and skid, there's very little rubbing on the road. If you drive reasonably slowly and carefully, your tyres will never slip and their grip will last a lot longer.

CHAPTER SIX

Sticky Stuff

In this chapter, we discover

> *How some glues are secretly powered by electricity.*
> *Why Post-it notes keep on sticking, again and again.*
> *How science can stop you sliding on ice.*
> *Why floors really are 'slippery when wet'.*

To stick or not to stick, that is the question – and, although it might sound unfamiliar, it's a question we ask in a thousand different ways, every single day. You might not use glues very often – you might not even have any at home – but *everything* you do, literally everything you do, involves things either sticking together or sliding straight past. With every breath you suck in, invisible gas has to slip through a maze of tubes, into your lungs, without getting stuck on the way. The same goes for the food and drink you gobble down. When you heave yourself upstairs, you can do so because you stick (very briefly) to the carpet or wooden boards beneath your feet. And that's all before we start to think about more obvious kinds of stickiness, like glues that keep stamps safely stuck on envelopes. Most of the time we notice the difference between sticking and sliding only when they fail: when sticky things slide or slippy things stick. 'Caution wet floor!' signs in supermarkets, dog ears of peeling wallpaper and beermats that cling to the bottom of your glass are typical reminders of how much we depend on getting the science of stickiness – stick or slip – right first time.

What makes one thing stick to another?

A magnet sticks to your fridge because an invisible force, magnetism, clamps metal to metal; there's no glue anywhere in sight. The same holds true for all kinds of stickiness and slippiness, whether there's glue there or not. Whenever things stick, there's a force holding them together; when they don't stick or slide past, the force is usually still there, but it's too small to bind.

Suppose you're putting up some particularly fine but rather heavy wallpaper, and it keeps peeling and curling back without sticking. What's going on? On the face of it, there's a simple fight between gravity (the weight of the paper, which makes it peel) and glue (making the force that sticks the paper to the wall). It sounds simple enough, but it's slightly more complex in practice because there are actually three different sticking forces involved. The paste (the glue) has to stick to the paper. Then the other side of the glue has to stick to the wall. Also, much less obviously, the paste has to stick to itself. When the paper peels off the wall, you might find it still has some of the paste on the back of it, while the rest of the paste has come off and stayed behind. What's happening here is that the paste is sticking to the paper and the wall but not to itself. Another example of this happens when you make yourself a jam sandwich, then peel the bread apart. You'll probably find jam stuck to both pieces of bread. When you pull the sandwich apart, the glue − the jam − fails first because it can't stick to itself with as much force as it can stick to the bread on either side.

Cohesion and adhesion

Looking at things more closely, our three sticking forces are really just two: things can either stick to themselves or to other things. We call these two types of force **cohesive** and **adhesive** (or cohesion and adhesion). We refer to glues as 'adhesives' because we assume that their stickiness boils down to tough adhesive forces when we smear them on other things; but, in fact, every effective glue also requires powerful cohesive forces to ensure that it doesn't simply break apart in the middle. Glues might be more accurately called 'adhesive-cohesive-adhesives' to reflect the three different types of sticking each one requires.

Water is probably the most familiar and remarkable example of adhesive and cohesive forces working side by side. Seriously cohesive, it happily sticks to itself even when there are other things nearby it could grab hold of instead, which is why rain splashes down in drops, when lots of water molecules all clump together. That might make you wonder why a rainstorm doesn't fall as a single, terrifyingly gigantic drop – and the answer is that large drops are unstable. Collisions between falling drops and air brushing past them as they rain down continually make them break into smaller pieces, so they never get the chance to grow bigger than about 5 mm (0.2 in)[61]. Water is typically much more cohesive than adhesive, which explains why a drop of water sits in the palm of your hand or pearly raindrops perch happily on leaves without slicking out across them. On a wet day, when the rain pounds your windows, you'll notice that the droplets tend to snake down definite, streaky channels – almost like vertical drains on the glass – and that's also because water is cohesive. Each new droplet has more inclination to stick to the ones that are there already than to other parts of the glass that aren't yet wet. Water is so good at sticking to itself and so bad at sticking to other things that we have to use detergents (surfactants) to make it spread out properly and wet things completely. We return to this in Chapter 17, when we look at why water is good at getting things clean.

Stickiness is mostly about adhesion, so let's focus on that now. Broadly speaking, we can divide it into three categories. There's permanent sticking (which is what we get from adhesives), temporary sticking (when we walk across the floor or a fly crawls up the wall), and downright non-sticking or slipping (when a lubricated razor glides across your soapy chin, or you slip and slide on snow and ice). Although it might seem as though these are completely different, they're all based on the short-range forces between two very close surfaces, and they have a great deal in common.

Permanent sticking

To stick one thing permanently to another, you have to make a super-strong physical or chemical bond between them. Suppose you weld a new chunk of metal to your car to replace a bit that's gone rusty. You're actually melting the two metals so that their atomic structure fuses together and they become one. That's not exactly a physical bond: you start out with two metals and end up with one. Not really what you'd call sticking, is it?

But what if you go to a cobbler's shop to get a new rubber sole stuck on your shoe; how is that different? It's not possible to melt the rubber onto the shoe without spoiling one or the other, so cobblers use glue as an intermediary: they smear a tough glue on the two surfaces and press them together. What we get is a 'bridge' linking the two different materials, sole and shoe, together. Exactly how they stick depends on the type of glue involved and what it's sticking to on either side. Some glues worm their way into the structure of both surfaces, hooking around the pores or nobbles of each and making a strong, physical connection between them. Some cause reactions to take place at the two surfaces they touch, creating a strong chemical bridge. Others are partly **adsorbed** (spread across each surface) and make a bond by creating tiny electrostatic forces between them. Still others work by swapping molecules and merging, which is called **diffusion**. These four different processes are illustrated below.

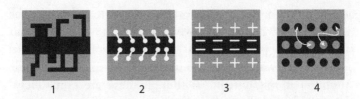

▲ **Four ways in which a glue (*black*) can stick together other materials (*grey*).** 1. Physical. *The glue can seep into gaps in the materials it's sticking, making a mechanical 'lock' between the two.* 2. Chemical. *The glue can create chemical bonds (white) spanning the joint, for example, by creating a new chemical compound where the different materials meet.* 3. Electrostatic. *Atoms in the glue pull electrons more or less than atoms in the other material, creating billions of tiny clinging forces that add up to an awful lot of stickiness.* 4. Diffusion. *The glue and the material it's sticking to swap their molecules, mingling, merging and binding together in a kind of atomic confusion.*

It's electric!

To me, by far the most interesting of these processes is the third one. Even if they smell disgusting and their tubes are covered in convoluted chemical names, some glues stick things together with *electricity*. That sounds extraordinary until you stop to think about static electricity and how it can stick things together like magnetism. If you rub a balloon on your pullover, you can make it stick to yourself or to the wall – yet there's no glue anywhere to be seen.

How does this work? All materials are built from atoms – and atoms, you'll remember, are crammed full of those smaller bits we call protons, neutrons and electrons. Protons have a tiny positive electric charge while electrons have a tiny negative charge. Overall, atoms have no electric charge because they have the same number of protons and electrons, and the two cancel out. But not all atoms are equal: some are greedier than others. If you put two dissimilar materials near one another and repeatedly rub them together, the atoms in one can 'mug' electrons from the atoms in the other. That's what happens when you rub a balloon on your pullover. The mugger (your pullover – and the material stealing the electrons) becomes negatively charged, while the unlucky victim (the balloon – with fewer electrons) is left positively charged. Like opposite ends of magnets, opposite electric charges attract. So the balloon sticks to your pullover.

A similar thing happens with some kinds of glue. When you bring it up close to another material, each microscopic molecule in the glue might be attracting or repelling electrons from a nearby molecule in the surface it's sticking to, creating one tiny electrical bond. So, to return to the cobbler's shop, the glue on your shoe might become slightly positively charged (say) while the surface of your shoe might become slightly negatively charged – and the two stick. These forces work over incredibly small distances, so they can be immensely powerful[62]. By 'short distances' we're talking about billionths of a metre. If, like me, you find things that small impossibly hard to visualise, imagine we scaled up this distance (a billionth of a metre) by about 100,000 times so that it was the width of a human hair (about a tenth of a millimetre) and we could see it with our eyes. Then a human hair, scaled the same way, would grow to about 10 m (33 ft) across – which is roughly the length of two cars parked bumper to bumper. It's not just the distance that counts. Because the sticking forces happen between every molecule in the glue and every molecule on the shoe, the effect is multiplied trillions of times. A very large drop of water weighing a tenth of a gram contains about 3,000,000,000, 000,000,000,000 molecules – three billion trillion – so that's very roughly how much the effect of one sticking molecule is scaled up in a water-based glue[63]. That's why, even using an apparently puny force like static electricity, a glue can still bond with amazing strength.

How much strength, exactly? A single square millimetre (that's 0.002 in^2) of superglue can hold the weight of a couple of bags of sugar. This sounds impressive, but it pales into insignificance compared to the world's stickiest thing, a freshwater bacteria called *Caulobacter crescentus*, which is about three times stronger (making an incredible sticking force of 70 newtons per square millimetre)[64]. This amazing natural super-stick offers the promise of better, natural medical adhesives in the future. In the meantime, for most of us, superglue is more than sticky enough. It's been capturing our imagination since the late 1950s, when one of the chemists who developed it, Professor Vernon Krieble, used a single drop to lift a man off the floor on the popular TV game show *I've Got a Secret*[65].

NOTE TO SELF: INVENT BETTER GLUE

Wouldn't it be great if glue stuck when you wanted – and not when you didn't? That's the secret thought that rocketed Post-it notes to astonishing worldwide popularity in the 1980s. When 3M chemist Spencer Silver and colleagues filed their patent for a 'Removable Pressure-sensitive Adhesive Sheet Material' in February 1973, no one could have guessed that they'd come up with one of the most useful everyday ideas of modern times[66]. In this very dry technical document, Silver runs through the disadvantages of conventional sticky tape and how his novel chemical preparation gets around them. The brilliance of the idea was only spotted a couple of years later when one of Silver's 3M colleagues, Art Fry, found himself daydreaming as he lost his place in a hymn book. What if you could make a sticky-unsticky bookmark that peeled off without damaging the paper below? So the Post-it note was born.

How does it work? If you were gluing a piece of paper permanently into a book, you'd smear a layer of glue right across it and press it firmly and evenly. This would make the adhesive spread out into a smooth, continuous, very thin film that would be impossible to remove without damaging either the paper or the book. But a Post-it note is different. Instead of being spread uniformly, the sticky stuff is a plastic adhesive, technically called acrylate polymer, contained in 'micro-capsules' (a technical name for blobs) around 100 times bigger than those in conventional glues, which forms a relatively rough and uneven sticking surface[67]. When you press a Post-it note down onto a page, some of the capsules touch the page to make it stick – but not all of them. When you pull the note off, the unused capsules allow you to stick it straight back down somewhere else. Eventually, all the capsules clog with dust and dirt – and the note doesn't stick any more.

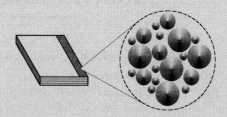

▲ **Sticky side up.** *Turn over a Post-it note and slide it under an electron microscope, and this is what you might see. The sticky part is covered in a scatter of microcapsules of different shapes and sizes. When you press down the note for the first time, the biggest capsules stick while the others remain unused. The second time around these large capsules are less sticky (because they've attracted dust and dirt), but the medium-sized capsules can stick instead. At the third attempt it's the turn of the smallest capsules – and so on, until there are no more sticky capsules left to do the job.*

▷

Temporary sticking

This 'electrical' model of how a glue works gives us a clue about our second kind of sticking: the friction that stops us slipping as we scoot across the floor. If it weren't for friction, walking would be impossible: every time you put your foot down, it would slide out from under you. Driving would be impossible too: a vehicle's wheels would spin around getting no traction and you'd go nowhere. Friction is a kind of temporary glue that sticks legs (and wheels) just long enough for them to shift to a new position, slightly further forwards – and that's the essence of how we move.

How friction works

Friction works in a similar way to the electrostatic glue we considered a moment ago. When two surfaces meet, the atoms in one are within striking distance (about five atom widths) of atoms in the other[68]. That's just enough to hold the surfaces briefly together. If friction works in the same way as glue, why isn't it a permanent sticking force? If you park a car on the street, why doesn't it stick there forever? How come you can ever drive it away?

It's all a matter of scale. Friction (low-powered gluing) and adhesion (high-powered gluing) differ in the size of the adhesive force they create. When you park your car the frictional force between the rubber tyres and the road is big enough to resist any force your car is ordinarily likely to encounter. Gravity (the car's weight) is not big enough to shift it, nor is any kind of pushing force you can apply with your own body. For all practical purposes a parked car is glued to the road. However, if you drive a truck or a tank at it, very slowly, you can easily push it out of the way. Similarly, there's going to be a limit to how steep a hill you can park a car on without it sliding down. At some critically steep angle, friction won't be able to balance the car's weight – and off it will go.

Gecko glue

Cars are huge and heavy things, although even toy cars (with their wheels somehow locked in place) won't stay still if you park them on steep slopes. But imagine a much smaller, lighter car with bigger and flatter tyres. Imagine if we made some kind of super-tyre with each ridge built from lots of small tyres and each one of those built from even smaller tyres. If we engineered it correctly, we could have billions of microscopically tiny tyres gripping the surface. If the car weren't too heavy, we could park it on the wall or even drive it upside down on the ceiling. What we'd have invented would have been the automotive equivalent of a gecko – the lizard version of Spiderman – that could clamber up walls. Geckos manage this incredible feat using *incredible feet*: each one has toes with tiny bristles, called setae, and each of those is covered in thousands of small hairs known as spatulae. Altogether, there are something like a billion hairs on each of a gecko's feet, creating an enormous surface area of *electrostatic* attraction. Geckos really do stick with electricity[69] – and that's why they can clamber up walls. If your hands and feet had as much clinging power as a gecko's, you could walk upside down on the ceiling with a 20-tonne rucksack on your back[70].

Friction breaks down eventually because you can always find a bigger force somewhere. That holds for *any* kind of glue, however permanent it seems to be. Apply a big enough force and you will break either the adhesive bonds between the glue and the surface, or the cohesive bonds within the glue itself. Or, if the glue is super-strong and it's applied to weak materials, the materials themselves might break in two, leaving the glue smirking up at you, happily intact.

Slipping

If gluing (permanent sticking) and friction (temporary sticking) are caused by forces, slipping is easy to understand as the absence of those forces. If you want to make something rough slide past something equally rough, what you need to do is minimise the frictional forces between the two surfaces where they come into contact. What's the best way to achieve this?

To make something like a floor really slippery, you have to coat it with a lubricant – and water will do nicely. It's even better when you add a little soap, so that the water forgets to clump together in droplets and spreads out evenly across the surface. There are two quite different reasons why a wet floor is so slippery. As we saw in Chapter 3, water is incompressible: you can't easily squash it into any less space and, because it's relatively dense and heavy, it doesn't move out of the way very quickly. If there's a wet patch on your laminate floor and you tread on it, it doesn't squirt out sideways or squash down like a sponge. Instead, for a time at least, you have a layer of water between your foot and the floor. Now the water might not compress, but it's a liquid, so it can move and flow relatively easily. As you tread on it, it forms a cushion between the rough surfaces, helping to reduce the friction between them.

But here's the clever bit. The water on the floor isn't a single layer like a block of wood: it's stacks of molecules, zillions high. Think of water on a floor as a whole series of layers stacked one on top of another. Each layer, called a lamina, can slide past the layers beneath it. You've probably seen this happening on a beach. When the sea is relatively calm, incoming waves wash in and break up the sand, while the waves that broke a short time before are racing back down the beach to the ocean beneath. You can see layers of water sliding past one another, some moving up the beach towards the sand and others, directly beneath them, sliding back down below. Water, in short, will happily slide over other water. So, when you put your foot on a wet and soapy patch on your freshly washed floor, you're applying a force to a whole stack of water layers, making them shear off sideways. Each layer moves a little bit, and you slide sideways

▲ **The science of slip.** *You slide on a wet floor because the water under your feet acts as a lubricant between your rough sole and the floor. Lubricants consist of multiple layers of fluid (massively exaggerated in scale in this figure) that slide past one another as you apply a sideways (shearing) force.*

and tumble to the floor. Put this idea to good use and you get the fun sport known as skimboarding, which involves surfing on a thin wooden board in the shallow, slippery water at the edge of a beach. It's an example of what's called **laminar flow**, which we return to in Chapter 15.

If nature gives us the stickiest substance in the world (that remarkably clingy river bacteria), what about going to the opposite extreme: what's the slipperiest thing we're ever likely to encounter? In everyday life, it's PTFE – the synthetic chemical better known as Teflon, the non-stick cookware coating that prevents omelettes from clinging to pans. In the natural world there's something even better: the lining of those carnivorous pitcher plants – a super-slippy water slide that helps nosy flies, spiders and frogs tumble to their end. Just like a soapy floor and a skimboarding beach, plants like this work by creating a slippery, intermediate water layer that has minimal friction[71].

Ice ice baby

In theory snow and ice shouldn't be slippery at all. Ice is a solid, the soles of your feet are solids and putting two solids into contact – such as a car tyre and a bit of tarmac – usually

produces enough friction to stop any movement. So why is
ice slippery? The standard explanation used to go something
like this: it's a general rule in science that if you squeeze
things, you raise their temperature. That's why bike pumps
get hot as you thrust them in and out, squeezing air into
your tyres. Theoretically, then, when you stand on ice, you
squash the uppermost layer, heat it up and melt it. So what
you get between your solid feet and the solid frozen ice
beneath is a layer of water, thick enough to act as a lubricant.
When you slide on ice, you're not sliding on the ice at all –
you're sliding on a thin layer of meltwater resting on the
surface of the ice. You can take this theory a bit further in
the case of wintery sports such as ice skating and curling. Ice
skates supposedly concentrate the pressure of your body on
a super-sharp blade, melting the ice much more effectively
right beneath your feet and providing just enough lubrication
to slide along at speed. Very little water needs to melt to
make this happen, and it refreezes quickly enough to prevent
the entire ice rink from turning into a lake[72].

Or that's how scientists used to explain it – even stellar
ones like Nobel-Prize winning physicist Richard Feynman[73].
Science, however, is a work in progress and always will be, as
we constantly strive to knock our spherical cows into better
shape. We now know that ice is even more complex than
previously thought. The real reason for its slipperiness has
nothing to do with the squashing pressure of a skater, which
simply isn't big enough to melt the ice into a lubricating
layer of water. Although we still don't have a complete
explanation, one currently favoured theory is that ice has a
built-in, liquid-like coating that increases in size the hotter
it gets[74]. Ice is slippery – because ice is slippery: it's a basic
feature of water, whether you're skating on it or not.

If you're out and about on the ice, what's the best way to
stay on your feet? You want to maximise friction, so rough
shoes with limpet-like grip are obviously a good start.
Slightly absorbent shoes, such as leather soles, sponge up

water and make it less likely to form a slippery lubricant layer underneath you. What about snowshoes? In theory, wide soles spread your weight over a much bigger area, reducing the pressure beneath your feet so you're less likely to sink into soft snow. But they're not so good on ice, where sinking is a good thing if it stops you sliding. The best option is probably crampons (snap-on ice studs), which concentrate your weight on a smaller area of contact. Like a mountain climber's pick-axe, the pointy prongs push straight through the lubricant layer and get a firm grip in the solid ice beneath.

Then again, you could just give in to the inevitability of science and jump onto your sledge.

CHAPTER SEVEN
The Inside Story

In this chapter, we discover

How to split an atom in your sitting room.

How many atoms it takes to power a light bulb.

Why salty fish and chips don't turn iron atoms to rust inside knives and forks.

'We're all the same on the inside' was the truth that inspired such heroes of the civil rights movement as Rosa Parks and Martin Luther King, but it's a fact of science as well as a tenet of social justice. And it goes well beyond most people's ideas of equality, because it applies to every single thing on Earth. Whether we're talking about human skin, cowhide, plastic shrink wrap or the bark of a tree, what's inside is atoms – and atoms matter if you really want to understand *matter*.

What makes someone 'tick'? Unless you know them well, you can never quite predict how they'll act in a certain situation. How they behave on the outside depends entirely on what they're like inside: where they were born, how they grew up and what they learned, who they've mixed with, what they've seen and done. Whether we're talking about crystal glass decanters and ivory chopsticks, or fibreglass surfboards and sheepskin rugs, the same holds true for all the little knick-knacks in your home: what they're like on the outside depends on their secret inner stories. What makes materials tick are the atoms and molecules jiggling inside them and the way they're locked together.

If you're relaxing at home at the moment, you're probably surrounded by about a dozen types of material: paper, card, wood, plastics of various kinds, numerous different metals, glass, ceramics (china and other pottery), glues, textiles such as cotton, wool and polyester – to say nothing of the living, natural materials inside the rubber plant on the window-sill and the sausage roll on your plate. Unlike people, who bumble their way through their careers mostly by accident, everyday materials are extremely well matched to the things we use them for. The famous Peter Principle of management science claims that people are promoted until they reach 'the level of their incompetence', but materials like plastic and glass soar to a peak of perfection without even trying; you get the sense that if an antique chair could talk, the oak inside would purr, like a village vicar clasping a well-worn bible, about a blissful life passed in a perfect vocation.

Most materials are so well matched to the jobs they do that we barely even notice them. It takes the artful shock of a fur-lined teacup (Méret Oppenheim) or a limp clock baking in the sun (Salvador Dalí) to provoke us into remembering how good we are at matching useful materials to the things we need them to do. This, of course, is the ultimate secret of human civilisation. From Stone-Age clubs to 'Gorilla glass' iPhones, progress has leaped through the centuries because people have constantly found new ways to put better materials to better use.

Lego of life

Peer through a powerful enough microscope and whatever you look at, on a sweeping encyclopaedic scale from aardvarks to zoetropes, will show up as nothing more than a bunch of atoms. As you might expect, airy things like helium gas (the stuff that swoops birthday balloons high in the sky) are built from the lightest and simplest atoms, while weighty things like uranium (nuclear fuel) are stuffed full of big and hefty ones. Absolutely everything on Earth is built from about a hundred different kinds of these invisibly small ingredients, which are very much the Lego blocks of life, but you can make virtually anything from just a handful of atoms. With only carbon, hydrogen, nitrogen and oxygen, you've got the raw material to construct most living things and an awful lot of non-living ones as well (like most plants, most plastics are built from these four atoms too).

We all know that atoms are small, but how small exactly? Stretch a tape measure across a typical atom (don't worry, for the moment, what that tape measure is made of or how you're going to hold it) and it'll measure about 0.25 nanometres (0.25 billionths of a metre). It's hard to visualise such a tiny slice of size, but let's try. If you take a millimetre (a thousandth of a metre) and divide it into a million equal pieces, each of those is a nanometre, and if you split that four times, you've got the width of an atom. Still can't picture it? Try this. There are about seven billion of us on Earth so, if people were atoms, and we stacked the planet's entire population on one another's heads, they'd make a pile roughly as tall as a typical adult. *That's* how small atoms are.

We meet again, Mr Bond

What makes one material different from another isn't simply the atoms stuffed inside it but the way they're connected – or, as chemists like to say, *bonded* – together. The most obvious illustration of this comes from the most familiar and essential material on Earth: water. Just as atoms are the simple units of chemical elements (like silver and gold), molecules are the basic building blocks of more complex things. Put different atoms together and molecules are what you get. So if you join two hydrogen atoms to one oxygen atom, you get a water molecule (H_2O). If you scoop several billion trillion water molecules onto the end of your finger, what you'll have is a single tiny droplet of water. Stick those molecules in a glass in your fridge and you'll get a tiny splinter of ice. Tip them into your kettle and flick on the switch and you won't have water or ice but a cloudy puff of steam. Exactly the same atoms, exactly the same molecules – but completely *different* materials.

Why? In ice, the molecules are jammed close together and bound reasonably tightly, although they still jiggle about to a certain extent. In warm water there's slightly more space between them and they can slip and slide past one another, which is why water flows and pours. In steam, the molecules are blown much more widely apart and zip back and forth like atomic boy racers. Or think of it this way: a block of ice is quite happy to sit there jingling and jangling in your Coke; a bucket of water, hurled horizontally, will dash across your floor but get no further than the walls; but a cloud of steam from a boiling saucepan will rapidly rush to fill your whole kitchen. Almost fill a litre (1.75-pint) bottle with water and freeze it solid in your fridge and it will bulge out somewhat, because water expands when it cools. Thaw it back to water, stick it in a saucepan, boil it dry and you'll be able to fill a large wardrobe with the steam you make. That's because steam occupies about 1,600 times as much space as the same mass of water. This is all you need to know to explain why boiling a kettle for even a few seconds too long makes your kitchen steam up like a laundry. Run the same magic trick in reverse on a typical cloud, a giant waft of cold water vapour, and you'll have enough liquid to fill an Olympic swimming pool[75].

How do we know about atoms?

If we can't see atoms, how do we know they're there? How can I persuade you that there are atoms under the floorboards any more successfully than I can convince you that there are mice scrabbling around in the attic? What makes the science of atomic physics any different from the leap of faith believers make when they embrace their gods? Science, unlike religion, is fuelled by evidence – and it's the accumulation of many bits of compelling evidence that convinces us atoms truly are the building blocks of our world. The evidence has been stacking up for about 2,500 years. Even the ancient Greeks believed in atoms; the word atom, meaning 'un-split', was coined by Democritus, the first person to populate our world with these 'invisible indivisibles'. This was a remarkable intellectual achievement, considering that there was little obvious evidence for the theory at the time.

The modern evidence for atoms comes in four main forms, from chemistry, electricity, radioactivity and atom splitting. First, there's simple chemistry. If you spot that a couple of different gases (call them hydrogen and oxygen) can fuse to make a liquid (water), and always do so in the ratio of 2:1 (two hydrogens for every oxygen), you can figure out a relationship between the basic chemical elements and the atoms from which they're made. If you mix a squidgy grey metal (call it sodium) with a nasty toxic gas (call it chlorine), you can make yourself some rather tasty table salt (sodium chloride) and sprinkle it safely on your food. Since sodium and chlorine join in the ratio 1:1, that clues you in on the relationship between two more chemical elements made from two different atoms. If you observe hundreds of different chemical reactions, you'll eventually figure out – as English chemist John Dalton did in 1803 – that the basic chemical ingredients always combine in simple multiples of one another: one to one, two to one, three to two or whatever it might be. Arrange the ingredients in a systematic order and you'll end up with a kind of two-dimensional cooking cupboard called the Periodic Table. That's the complete list of chemical ingredients we have at our disposal.

Simple chemical reactions suggest that atoms must exist, but take us little further. If things like water and salt are made from atoms, what are atoms made of? According to the ancient Greeks it's a meaningless question because atoms are 'indivisibles' – and not made of anything meaningful at all. That seemed a perfectly reasonable conclusion until the 19th century, when scientists such as Ampère, Volta and Faraday teased out the mysteries of electricity. In 1897, when Englishman J. J. Thomson discovered a fundamental packet of electrical energy, now known as the electron, a second piece of evidence emerged, confirming not just the existence of atoms but a whole, dazzling 'subatomic' world inside them[76]. Luckily, Thomson hadn't been put off by a cousin of his who asked him, as a schoolboy, what he wanted to do when he grew up. 'Original research,' Thomson had replied. 'Don't be such a little prig,' he was told[77]. What if he'd abandoned his dream at that point and simply become a doctor or an accountant? What if the electron hadn't been discovered for several more decades? Would the entire 20th-century electronics and computing revolution have been postponed by 50 years or more? It's pointless – but wildly fascinating – to speculate.

Splitting the atom

An even more powerful piece of evidence that atoms had a secret inner story was emerging in France at roughly the same time. In 1896 Henri Becquerel discovered that uranium spontaneously spewed out penetrating particles of radiation, similar to X-rays (which had recently been discovered), but much more powerful. Radioactivity, as we now call it, is easy to understand if you assume that giant atoms are crammed full of smaller particles. Some of these monstrous atoms exist in excitable, unstable forms, known as isotopes, that are itching to settle into more stable states – and they do that by flinging out all the bits they don't want or need.

Marie Curie (and her husband Pierre) took Becquerel's work a step further. Although it cost Curie her life, the sacrifice was a noble one: radiation treatment for illnesses such as cancer, directly inspired by her work, has saved countless lives ever since[78]. Scientific discoveries often seem glamorous in retrospect – Curie's inspired a 1943 Hollywood film starring Greer Garson and Walter Pidgeon – although the day-to-day grind of most laboratory research is anything but glamorous. We remember the leaps of insight and (occasionally) the science heroes behind them, but we tend to forget the numbing tedium that goes on behind the scenes.

Marie Curie's discovery involved repeating 5,677 experiments over four years, boiling some 8 tonnes of pitchblende (uranium ore, a German name that translates loosely as 'bad luck mineral') in a bubbling cauldron to produce, in the end, about 1 g (0.04 oz) of a useful radium compound[79].

Hindsight makes us shudder at Curie's dangerous adventure, because we know how her story ended. It's part of a long-running narrative that still prevents us from trusting scientists on the issue of nuclear power. Mushroom-cloud memories of Hiroshima have a very long half-life, while catastrophic accidents, at Chernobyl in 1986 and Fukushima in 2011, have fostered a deep suspicion of nuclear energy – even though natural radioactivity causes far more deaths from cancer[80]. But Curie's pioneer spirit still pops up from time to time. An entertaining issue of *Popular Science* magazine from July 1955 describes amateur uranium prospectors, working under cover of darkness with Geiger counters bought from mail-order catalogues, motivated by 'fat bonuses' from the Atomic Energy Commission:

A mother and son bought a short-wave ultraviolet lamp, read up on mineral prospecting, and recently spent evenings roaming the hills . . . The news spread and, within a month, hundreds of claims were staked.[81]

In February 2014, 13-year-old English schoolboy Jamie Edwards became the youngest person ever to achieve nuclear fusion (joining small atoms together to make larger ones); he'd previously saved up his Christmas money to buy a Geiger counter[82]. J. J. Thomson would have been proud.

Unstable atoms, such as radioactive uranium, break up to form more stable ones, but that's not the only way in which atoms can change. The fourth and final piece of evidence confirmed not just that atoms exist and that they contain smaller particles such as electrons, but their exact inner structure. Although New Zealander Ernest Rutherford is widely remembered for 'splitting the atom', the honour should really be shared between all those scientists who toyed and tinkered with bits of atoms in the first 20 years of the 20th century.

Rutherford gets the credit for the most famous experiment of them all, which was carried out in Manchester in 1910 by his two junior associates Hans Geiger and Ernest Marsden. Firing positively charged chunks of helium atoms at gold foil, they found that most passed through harmlessly, while some (roughly one in every 8,000) bent off at absurd angles and a few bounced right back the way they'd come. Rutherford was staggered, famously noting that it was 'as if you fired a 15-inch shell at a piece of tissue paper and it came back and hit you'. The explanation for this, to a hindsight generation, now seems obvious. The positively charged heliums had scored a direct hit on the positively charged central nucleus (middle) of the gold atoms, causing them to be repelled ('scattered', as the physicists put it) like the north poles of two

▼ **Splitting the atom, old style.** *Ernest Rutherford fired alpha particles (chunks of helium atoms) at gold foil and watched what happened next. Most of the alpha particles zapped straight through (1), relatively undisturbed. A few were bent through very large angles (2). One or two bounced straight back in the direction they'd come from (3). From this, Rutherford realised that gold atoms were made from a central nucleus surrounded mostly by empty space, dotted with electrons here and there, and calculated the size of the gold nucleus with reasonable accuracy.*

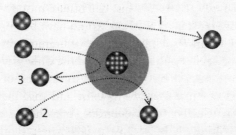

▲ **Splitting the atom, new style.** *When protons smash together in CERN's Large Hadron Collider, more than 100 other particles are produced in the collision, leaving tracks that show up as individual lines. (Artwork by Lucas Taylor. Copyright © CERN 2010)*[83].

magnets. Rutherford's neat experiment settled the outstanding mystery of what atoms are like inside. They are mostly empty space, with much of their mass positively charged and packed in the nucleus, while their negatively charged electrons 'spin' in a hazy cloud of nothingness all around it.

These days, atom splitting is old hat. The modern-day descendants of Rutherford's 'particle accelerator' have split atoms into particles and those particles into even smaller ones. We now know there are dozens of subatomic particles, from old chestnuts like protons and neutrons right down to that brand spanking new (and most elusive) Higgs boson, the particle that scientists spent decades and billions of euros hunting down in the giant circular atom smasher near Geneva known as the Large Hadron Collider (LHC)[84].

Not that you need either time or money to smash atoms. Until relatively recently most of us were doing it in our sitting rooms on a nightly basis. Old-fashioned cathode-ray tube (CRT) televisions work by 'boiling' metal heating elements so that they release electrons (historically called 'cathode rays'), zapping them down long glass tubes, then steering them with magnets so that they smash into the phosphor screen on the front, tracing out pictures on the screen.

WHAT FUELS FUEL? THE ENERGY
STUFFED IN STUFF

It takes *real* effort to bring atoms together – and the closer you push them, the harder it gets. Squeeze steam tightly enough and you'll get water. Squash it even more and you'll make ice. In exactly the same way, press a few gazillion carbon, hydrogen and oxygen atoms together and you can make yourself some very handy petrol, coal or firewood. The energy locked in these fuels that you release when you burn them is the very same energy that went into forcing the atoms together to build their molecules (hydrocarbons) in the first place. If we had to make fuels by hand, there would be no point in doing so because we'd only ever get back the energy we supplied ourselves. Fortunately, fuels are made by nature – ultimately by the Sun and the natural pressure and heat within Earth – so we get energy out of them without having to put it in in the first place.

What holds for molecules holds just as well for atoms. In theory you can make an atom by ramming together the tiny bits it's made from (protons, neutrons and electrons). Although you need a huge amount of energy to make it happen, you can get it back when you're done. We call this process **nuclear fusion** (because atoms are being made by joining bits together). By the same token, you can smash apart atoms to release blistering amounts of power. No one realised you could do this until the early 20th century. Albert Einstein gave the first real clue in 1905 when he came up with his infamous equation $E=mc^2$. The speed of light (c) is a very large number (300,000,000) and c^2 (c times c) is even bigger (90,000,000,000,000,000,000), so even a tiny amount of mass (m) will produce a giant amount of energy (E). It all sounds a bit dry and theoretical until you remember the thumping practical consequences: the first most people knew about energy made from atoms was when two relatively small (3.3-m or 10-ft long) atom bombs

▷

obliterated Hiroshima and Nagasaki in 1945. This alternative way of releasing atomic power is called **nuclear fission** (because the atoms are being broken apart). Half a century later most of us will have used, at some point in our lives, electricity made by splitting atoms in a nuclear power station. What an amazing thought.

How many atoms does it take to power a light bulb?

Suppose we take 1 g (0.04 oz) of uranium (the heavy element used as fuel in most nuclear power stations) and split all its atoms to release energy. If we could do that every second, we'd generate 100 gigawatts of power. Now a hefty nuclear power station only produces about 2 gigawatts, so what we've got from a mere smidgen of uranium is equivalent to the output from about 50 plants[85]. That's because, in practice, nuclear plants use relatively little fuel, relatively slowly. Looking at the same numbers from the opposite direction, it means that if you want to power a 10-watt low-energy lamp, you need to split about 300 billion uranium atoms every second.

Big numbers, but no big surprise. Atoms, as we knew all along, are very tiny things.

What makes different materials so different?

With the exception of drinks, a few foods, and handy things like washing-up liquid, most of the materials we use around the home are solids. It's the scientific equivalent of interior design – the arrangement of atoms and molecules inside them and the way they're bonded together – that makes each of these materials behave differently from the others. So it's simple inner science that explains why you put glass in your windows and bricks in your wall and not the other way around. Open up any material you can think of, poke around a bit among the molecules and atoms, and you'll spot things that make it interesting and unique, perfect for some jobs and pants for others.

Let's take a quick peek inside a few common materials and see exactly what makes them tick.

Marvellous metals

Iron is just about the simplest material you'll ever find. Opening it up to peer at its inner structure is a bit like lifting the lid on a cardboard box filled with hundreds of equally sized marbles. Firm and round, each marble represents one iron atom; the atoms pack tightly in rows with one layer on top of another[86]. So far so good, but can this explain how an iron bar works? Iron is relatively hard (more so than chalk or cheese) because the atoms are packed closely enough to cling to one another. Yet it's also relatively soft (softer than steel or diamond). You can bash it with a hammer to knock it into better shape, and that's possible because the layers of atoms slide happily past one another. Unlike glass, another hard material, iron bends when you shape it because the atoms don't mind moving; glass shatters because the atoms can't move into new places without the entire structure falling apart (glass is explored in more detail in the next chapter.)

Iron conducts electricity reasonably well because the electrons in all those close-packed atoms merge together into a kind of soupy sea that sloshes back and forth through the entire structure, carrying electricity from one side to another. Heat travels through iron in a similar way, handed from atom to atom in an invisible game of 'Psssst, pass it on'. Iron glows red hot when you heat it enough because the atoms happily absorb heat energy but give it back out again as (red) light.

Even a material as simple as iron has science secrets hidden inside it. Heat it hot enough, add a small amount of aluminium and, when the whole thing cools back down again, you'll find that the aluminium has dissolved inside the iron, with the aluminium atoms incorporated perfectly inside the iron crystal structure. One metal really can dissolve another one – making what's called a solid solution.

Allies in alloys

Metals are pretty simple: often simply good, sometimes simply bad. One of the problems with iron, for example, is that it's relatively weak when you pull or bend it ('in tension', as engineers say). But if you add carbon to iron, so the tiny carbon atoms hide between the bigger iron ones, you get a much stronger material. This is how we make cast iron, wrought iron and steel, all of which are iron **alloys** (mixtures of a metal with one or more other substances). Steel is stronger than iron because those tiny 'interstitial' (intermingled) carbon atoms prevent the iron atoms from shuffling about so much.

Another big problem with iron is that it rusts very easily. One option is to paint it – over and over again in the case of a giant iron structure like the Forth Bridge. A better solution is to add some chromium to your iron-carbon mix, making the even more sophisticated alloy that we call stainless steel. You might wonder why stainless steel knives and forks don't rust when they're up to 90 per cent iron and spend half their lives mired up in noxiously rust-inducing substances like salty fish and chips. The answer is that the chromium atoms inside react with oxygen in the air to form a thin outer skin of chromium oxide, which prevents more oxygen and water from penetrating the vulnerable iron inside.

Perfect plastic

We've come to think of plastic as cheap, colourful and disposable – and that's what the word means to most of us. However, it's more accurate to think of *plastics* (because there are dozens if not hundreds of different ones) that are *plastic* (inherently flexible and useful in many different ways).

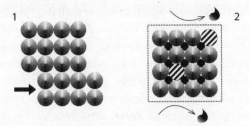

▲ **Iron in action.** *Iron is built from layers of atoms. You can bend and shape iron because the layers will slide past one another (albeit with some difficulty) (1). Stainless steel is harder than iron (2) because carbon atoms (small black circles) packed in the empty spaces make it harder for the iron atoms to move. Chromium atoms added to the mix (striped circles) react with oxygen in the air to form a protective outer coating of oxide (dotted lines) that stops most water from penetrating the structure. In alloys such as steel, iron gets by with a little help from its (atomic) friends.*

It's easy to manufacture plastic things so they look like metals, wood, glass, cotton or virtually any other material you can think of, but the resemblance is only skin deep. Inside, plastics are quite different from other materials.

Where a block of iron is built from iron atoms, a chunk of plastic isn't built from plastic atoms. Plastics are usually made from giant molecules called **polymers**, each typically based on the atoms carbon, hydrogen, oxygen and nitrogen. A polymer is made from endless repetitions of a smaller molecule called a **monomer** and often looks like a big long chain. If you imagine that a coal train is a polymer, the monomers are the individual, identical trucks hooked behind the engine. Though flexible, plastics are remarkably robust and resilient.

If you've ever taken part in a beach clean, you'll know that plastic represents by far the biggest quantity of refuse washed up from the oceans – and some of it can be astonishingly long-lived. It's estimated that typical plastics survive in the environment for up to 500 years. That's partly because their long-chain polymers are so resistant to attack by air, water, light, heat and all the other things that break up simpler materials such as paper and wood. It's also partly because – with the exception of a few peckish bacteria – nothing really eats them or has learned how to digest them[87]. The kinds of plastic that are now so ubiquitous haven't even been around for a century[88], so it's interesting to speculate what a world choked with indestructible plastics might look like in 500 years' time. It's hard for us to imagine a timescale like that, but worth bearing in mind that if Henry VIII had had plastics, we might still be finding his old toothbrushes in archaeological digs today.

Despite their long life, many plastics are soft and flexible because the polymer chains are held together only by fairly weak bonds. Unlike in metals, where there's a sea of free

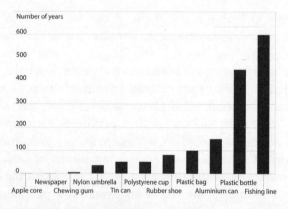

▲ **The long life of litter.** *It's no real surprise that plastics linger for longer in the environment than natural materials. What is surprising is how long they survive – a good few hundred years. What governs how long a material lasts is how well natural things such as sunlight, water, heat and bacteria can break apart its internal structure, turning it back into more benign things[89].*

electrons to carry electricity and heat, the electrons in plastics are all safely snuggled up inside their atoms, so heat and electricity don't surge through plastic materials anything like so well. That's not to say that all plastics are like this – and they're certainly not all soft and weak. Kevlar, a close relative of nylon, is built from microscopic, rod-like fibres that are jammed tightly together and pointing in the same direction like matches lined up in a box. That makes it, weight for weight, about five times stronger than steel. Stitch 30 layers of Kevlar together and you get a kind of super-thick 'plastic bag' that can stop a handgun bullet travelling at more than 1,500 km/h (930 mph)[90].

CHAPTER EIGHT
Amazing Glazing

In this chapter, we discover

> *Why you can see through thick glass (but not thin metal).*
> *How windows can clean themselves with nothing more than sun and rain.*
> *Why glass is so much heavier than we expect.*
> *How to stop bullets with a glass and plastic sandwich.*

Strip humans back to their essence – shrug off a few thousand years of culture and invention, and what passes, these days, for civilised behaviour – and we're little more than hunter-gathering creatures of the bush, no different from moles, meerkats or snuffling, shuffling aardvarks. But give us a rainy Sunday afternoon and there's nowhere we'd rather be than indoors, snuggled on the right side of the dripping windows with a book, a black-and-white film or some red wine and laughter. Most animals make shelters of some kind, but what marks out humans is our ability to create buildings that blend the best of both worlds. We might be curled up indoors but, with a few strategically inserted panes of glass, we can be outdoors at the same time. Metals and plastics, like the ones we looked at in the previous chapter, are humdrum materials with a million mundane uses. Glass, on the other hand, is much more cunning: what you *don't see* is what you get.

This *transparently* ingenious stuff is one of our oldest engineered materials: they were using it in ancient Mesopotamia something like 5,000 years ago, albeit to make sparkling, coloured, jewellery beads rather than vertical sheets of nothing[91]. Glass windows date back to at least

Roman times[92], and you might think that was both the beginning and the end of a very short story: what more is there to say about something invisible? In fact, people are still developing crafty new kinds of glass to this day. How about windows that scrub themselves automatically, or lighten and darken at the flick of a switch? What about the super-tough glass that fronts your mobile phone or tablet computer? Not so long ago people would have been scared stiff at the idea of balancing glass slabs in their laps; now we don't give it a second thought. Glazing is truly amazing – and it's all thanks to science.

Boiled beach

Here's a simple recipe for glass. Nip down to a beach with a bucket and spade, and shovel yourself up some sand. Build yourself a roaring fire. Tip the sand in, boil it to a broth, cool it quickly and – hey presto – what you'll get is glass. You'll need to add a few more ingredients to refine the quality (things like sodium and calcium carbonate make the process easier, and tossing in bits of metals such as selenium, iron or copper will give the glass a nice hint of pink or greeny-blue)[93]. In essence, however, that's all there is to glass: boiled beach, best served cold. When J. Robert Oppenheimer and his colleagues built the first atomic bomb in the New Mexico desert, on 16 July 1945, they tested this recipe with impressive results. Their sky-high cookery instantly turned a ¾-km (½-mile) wide circle of sand, directly beneath the bomb tower, into radioactive green glass[94].

Now if you heat ice to make water and cool it down again, what you get – save any steam you lose in the process – is more or less what you started with: water. The same isn't true of sand. If you heat sand into a liquid, then cool it back again, you don't get the solid you had originally. Boiled up, the atoms in the solid move apart and jiggle around: you can pour molten sand, blow it and tip it into moulds to make panes of glass. But when it cools, the atoms don't settle into the nice neat arrangement you find in a typical solid. What you get instead is a kind of random, ad-hoc structure somewhere between the chaos of a liquid and the order of a solid[95]. We call it an **amorphous solid** or, sometimes, a semi-solid or a frozen liquid. You'll often read in books that glass is a liquid that's never quite set (which can be a little bit misleading, for reasons we'll come to shortly).

Locked in limbo

It's easier to picture what's happening in glass if you think of people. Lots of them. Imagine a few hundred city suits hurtling through the concourse of a train station at rush hour. Swap atoms for people and that's what a gas is like. Now imagine a crowd of people packed inside the entrance lobby of a theatre. They're much closer together, squeezing past on their way to the box office, cloakroom or bar, but still perfectly free to move, albeit in a slow and somewhat sluggish kind of way. That's the human equivalent of a liquid. Stick the people in rows and columns, like ranks and files of soldiers, and you have a solid. They can still move a little, but if they're packed that tightly it's impossible for anyone to escape, or for the structure as a whole to change very much at all.

So what is an amorphous solid? What if you woke a barracks full of soldiers at 4.00 a.m. and gave them exactly a minute to get on parade. Amid all the shouting and darting, the half pulled-on jackets and quarter-woken faces, suppose you shouted 'Stop!' at exactly 60 seconds, forcing them to freeze. What you'd get would be a lot of people frozen in between order and chaos: a few hints of getting it together; a lot of people rushing in vaguely the right direction; much more sign of solid than liquid – but still far from the crisp, neat, *crystalline* arrangement you'd get if you gave people more time. That's what an amorphous solid is: a rough-and-ready attempt at a solid. Glass is by no means the only one.

▲ **Structure of glass versus typical solids.** *A typical solid (1) has a regular inner structure. We call it a crystalline, or crystal, structure, even though the solid, which might be something like a metal, bears no resemblance to the sparkling chunks of quartz you might see in shops. Glass (2) has a more mixed-up, amorphous structure lacking any obvious pattern or order.*

You can make amorphous ice, for example, if you cool water extremely quickly. Although we don't see it in everyday life, it's quite common in the wider Universe as a kind of cosmic frost: comets are mostly made of it[96].

While it's correct to say that glass is somewhere between a solid and a liquid, that doesn't mean it's still in the process of becoming a solid or will eventually solidify completely; it's as solid as it's ever going to get. It's also untrue to say that glass, being semi-solid, is like a liquid that flows in super-slow motion. One often-repeated myth in children's science books is that ancient windows are thicker at the bottom than the top because the semi-solid glass is very slowly flowing downwards. That turns out to be untrue; experts believe the real cause is that glass was made by a traditional (Crown-glass) process that left it thicker in some places than others, and it was generally then fitted in frames with its fattest edge facing down[97].

Why do windows smash and shatter?
Glass is nothing if not histrionic. 'Look at me! Look at me!' it screams – but, at the first sign of trouble, it cracks, splinters and shatters to smithereens. Most solids don't do this or

anything like it. Bang your fist on a desk and the whole thing doesn't shatter beneath you. Back your car into a wall and you might escape with a scratch or two or a bashed bumper, but you won't be left sitting in a pile of dust. So what's different about glass?

As we discovered in the previous chapter, it all comes down to the *inside story*. Metals are close-packed, crystalline solids in which atoms can slowly shift about, after a fashion: you can hammer metals into shape by bashing rows and columns of atoms into new positions. Since their atoms can move, they can readily soak up the energy your hammer blows supply. Glass, on the other hand, has an open, amorphous structure. The atoms aren't tightly packed in the same way but more loosely, randomly linked. Fire a bullet through glass and the atoms have no quick and easy way to rearrange themselves, and no way to absorb or dissipate the bullet's energy, so the entire structure collapses. Even the slightest bit of stress makes them crack.

'Bulletproof glass' is actually a misnomer: there is no such thing. The stuff sold under that name is neither bulletproof nor glass. It's bullet-*resistant* and made from a laminate (bonded layers) of glass and plastic. Fire a gun at this stuff and the bullet's energy spreads out and soaks through the layers, so it's rapidly dissipated. The plastic prevents the glass from shattering and helps to absorb the bullet's energy, so even if the bullet emerges from the other side it's not travelling fast enough to hurt you so much.

Blowing hot and cold
As all cooks know, you don't have to hit glass (or shoot bullets through it) to make it shatter: dash hot glass into cold water and – crack! – that'll do just nicely. Again, the problem is that the random, amorphous structure of glass cannot rearrange itself quickly enough to dissipate energy, which is all heat is, really. Stick a wine glass in a bowl of hot water and the

amorphous structure soaks up quite a bit of heat energy, with different bits of it expanding by different amounts as the heated atoms inside it jolt around. Cool it down quickly and the same atoms slow suddenly but can't quite realign and return to their original positions. The slightest flaw shoots through the structure as a crack[98] – and a shatter.

Cooks avoid this problem by using heat-resistant borosilicate glass, best known under the trade name Pyrex. It's much like ordinary glass but with an added ingredient, typically 13 per cent boron oxide, which gives a characteristic 'blue-rinse' effect if you look very closely. We can think of it as a kind of 'glass alloy': the boron oxide makes it expand about a third as much as ordinary glass when it's heated (or cooled), so the amorphous, atomic structure shifts about less and is less likely to break apart[99].

Why is glass so heavy?

If you've ever watched shopfitters heaving plate-glass windows into position, this question might have crossed your mind. After thinking about this for many years, I've come to the conclusion that the answer is partly down to physics – but much more to psychology.

First, the physics: remember that glass is, at least in some sense, a solid. If you have a plate-glass window that's, shall we say, 2 m across, 3 m down, and 2 cm thick (6.5 × 10 ft × ¾ in), the volume works out at 0.12 cubic metres or 120 litres. It's fairly easy to imagine how much 120 litre bottles full of water would weigh (roughly 120 kg/265 lb, about twice an average woman's weight) – and glass, remember, is a solid, not a liquid. If our window were made of stainless steel, it would weigh almost a tonne (because 1 cubic metre of steel weighs about 8 tonnes). Put glass in this kind of context and its weight isn't surprising at all. A plate-glass window weighs somewhere in between the weight of a block of ice and that of a lump of steel of identical volume, just as we'd expect[100]. In fact, our window would weigh about 300 kg (660 lb) – as much as 4–5 adult men.

What I think surprises us is the psychology: because glass is transparent and looks like nothing on Earth – indeed, like *nothing at all* – we think of it as emptiness and expect it to be light. But whether you think of it as a liquid, a solid or a lost limbo in between, it's still packed with zillions of atoms. You can't see it, but it's there nevertheless. That, of course, prompts the most important question of all.

Why can you see through windows?

Thousands of years before plastic – that cheap pretender – came along, glass was the ultimate in 'see-through'. Indeed, with the exception of water and a few natural materials (think dragonfly wings), it's still one of the only **transparent** materials there is. The transparency of glass obviously boils down to the fact that light can pass through it, but why can't it pass through other solids, such as metals?

Thickness doesn't come into it: thin sheets of paper are **translucent** (allowing light through but scattering and rearranging it in the process, so you can't really see what's there), but they're not transparent (letting light through in such a way that you can see clearly what's on the other side). Very thin tracing paper is transparent if you place it right next to something, but translucent if you hold it at a distance, because it partly reflects light (bouncing it back at a precise angle), partly transmits it and partly scatters it too (throwing it off at random angles, making accurate transmission impossible). Sheets of aluminium foil are less than two-tenths of a millimetre thick – thinner than most sheets of paper – but you certainly can't see through them.

Whether materials are transparent or opaque depends on what effect they have on light as it tries to wriggle through them. Metals readily soak up particles of incoming light, which are called photons, and quite often anything

similar – including things like X-rays (we'll peer at light in
more detail in Chapter 10). Their atoms are surrounded by a
sea of free electrons, which easily absorb the photons of
light and spew them out again, effectively catching the
photons like balls and tossing them back the way they came.
Shiny metals like aluminium and silver catch all kinds of
photons (light of all colours) and throw them back again,
which is why they make excellent mirrors. Coloured metals
such as copper or gold absorb some photons but reflect or
transmit the rest (copper reflects the reddish part of any
incoming light, but soaks up the rest of the spectrum from
yellow and green to blue and violet).

What's different about glass? Once again, it's the *inside
story*. In glass, all the electrons are fully occupied holding the
atoms of that weird amorphous structure together, and they
need much more energy to excite them. That means they
can't capture puny photons of visible light the way metals
can, so those ordinary photons simply wriggle in one side of

▼ **Why you can see through thick glass but not through thin
metal?** *When sunlight shines on copper (1), it excites electrons in the
metal atoms inside. The excited atoms are unstable and throw the light
back out again, although the light that leaves is redder, which is why
copper looks characteristically reddish-brown. Things are different in
glass (2). When ordinary light shines into it, it doesn't have enough
energy to excite the atoms. It simply passes them by and emerges more or
less unchanged on the other side (although its path is bent slightly or
'refracted' as it enters and exits).*

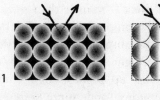 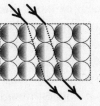

1   2

a piece of glass and out the other. The atoms in the glass barely even notice them. If we use ultraviolet light, however, its photons are more energetic than those in visible light and just right for glass to absorb. That's why glass generally looks opaque in pure ultraviolet light[101]. When you colour glass, by adding atoms of metals, you get the best of both worlds: light passes through, but the metal atoms inside change its colour on the way.

Better kinds of glass

A *solid* we can see *through*: glass is one of those everyday things that's so utterly amazing, so completely extraordinary, that we take it for granted and consider it mundane. We often don't even notice windows are there until passing, scared sparrows carelessly spatter them. That's not to say that glass is perfect – not by any means. It's brittle and fragile, flaunts dirt to an irritating degree, lets heat escape all too readily and (like the worst kind of snitch) shares our secrets with anyone who wants them. Fortunately, all these problems have intriguing scientific solutions; there really are better kinds of glass.

Stop that heat

Lounging by a window on a winter's night can be cold and draughty, but it's not just the leaky joints where the glass meets the frame that cause the problem: the glass itself lets heat escape. If it's not obvious why a solid (or pretend solid) like glass should let heat seep through, think about sitting around a camp fire – or, if you've never done that, in front of an open coal or wood fire. If you're fairly close, you'll feel the heat making your cheeks glow even though there's little more than air between the fire and your face. It's almost as though you're being warmed by invisible rays of heat – and, indeed, that's exactly what's happening.

Heat can hurtle through emptiness – even from the Sun to Earth, through the vast vacuum of space – in the form of

infrared radiation. Heat is not so different from light and gallops along at the same speed (300,000 km/186,000 miles per second). The only real difference between heat (infrared) and ordinary light is that the waves that carry heat are very slightly longer. If you take a rainbow that arcs from red (on the outside) to blue (in the centre), infrared perches just beyond the red, outside the colours you can see. If heat is similar to light and travels the same way, it should come as no surprise that it can shoot straight through a glass window: where light can go, heat can follow.

The solution is simple: make your windows into partial mirrors by smearing the glass with an ultra-thin film of metal or metal oxide (something like titanium dioxide will do nicely). Make the metal just a few atoms thick and enough light will still get through. But, on baking hot summer days, the almost invisible coating will reflect heat back outdoors, keeping your home reasonably cool inside. On chill winter nights, when your home is warmer than the freezing snowscape outside, any heat you make inside (with gas central heating or the buzzing red bars of electric fires) will hit the metal coating and bounce straight back, so it won't escape as readily.

Squeaky clean
Most people love sparkling windows, but not everyone shares my enthusiasm for wobbling up ladders with a sponge and a chamois leather. You might wonder why windows, with copious supplies of rain hitting them more often than not, don't stay clean all by themselves. But dirt builds up and clings to older dirt; there's only so much a gentle drizzle can do, poor thing.

Fortunately, there are now ingenious self-cleaning windows. Unlike car windscreens, which scrub themselves with a squirt of detergent and a lot of help from a squeaky mechanical wiper, self-cleaning windows work through chemistry, not engineering. Like heat-reflecting windows, self-cleaning ones have a built-in chemical coating of something like titanium dioxide. As sunburned sunbathers know only too well, sunlight has a harmful, invisible component called **ultraviolet**. It's the stuff that wrinkles your skin like overcooked Christmas turkey and, if you're really unlucky, gives you skin cancer. Ultraviolet is a bit like a blue version of infrared. It has a slighter shorter wavelength than ordinary blue light and perches, invisibly, on the inside of a rainbow. Think of infrared and ultraviolet as 'bookends' or 'hard shoulders' on the two extreme sides of the colourful spectrum we can see.

Harmful to skin it might well be, but ultraviolet light certainly has its uses. When it hits the titanium dioxide molecules on a self-cleaning window, it knocks electrons out of them. They crash into water molecules in the air and convert them into what are called hydroxyl radicals: H_2O water molecules split apart so one of the hydrogen atoms (H) shears away, leaving a manic OH, a hydroxyl radical, behind. These radicals work like detergents, chopping up dirt into much more manageable pieces. Next time it rains,

▼ **How a window can clean itself.** *When sunlight (1) falls on the titanium dioxide coating, it generates electrons (2). The electrons target water molecules in the air, knocking hydrogen atoms off them to convert them into active hydroxyl radicals (3). The hydroxyl radicals attack dirt (4), breaking it into smaller pieces. When it rains, the chunks of dirt slide naturally off the glass (5).*

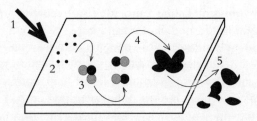

the fragments of dirt wash clean away. At least in theory. Shame about the insides of the windows, of course, which you still have to clean yourself.

Now you see it, now you don't

One of the most interesting – and in some ways baffling – things about a window is that it gives you a completely false impression of what's on the other side. If you're ambling down the street and you happen to peer through a shop front or the front window of someone's home, you can't actually see very much at all. If you're inside, looking out, and you see someone staring in at you from the pavement, you might reasonably conclude that people passing by can watch everything you're doing; that's why so many people use net curtains.

But this is completely wrong. And it's all to do with the fact that there is much less light inside than out. How much less? It obviously depends on the time of year, the time of day, the weather and where you are on Earth (which affects the apparent position of the Sun in the sky) but, broadly speaking, the average light brightness outside is about 2,000 times greater than it is indoors[102].

If you're inside there's tons of light outside – so a great deal of it is transmitted through your window and you can easily see what's going on. If you're outside there's lots of light around you, but there's very little light in the house and even less of it successfully sneaks through the windows. A piece of glass might look clear and transparent, but it doesn't let all light pass through: it reflects about 5–10 per cent of the light falling onto it[103]. So even though there's some light inside, not all of it can escape. That's why, perhaps surprisingly, windows are extremely good at preserving your privacy most of the time, whether net curtains are hanging there or not.

Of course at night the whole thing switches around. Once the Sun disappears and we're left with street lights and the milky glowing moon, there's almost no light outside: indoor light is about 500 times stronger than moonlight[104]. Indoors, with the lights switched on, there's suddenly light to spare. If you look out through your window in darkness, at a person looking in, you won't be able to see them, even though they can see you quite clearly. All you'll see, most likely, is your own reflection in the glass.

Look, no curtains

Most of us solve the problem of night-time privacy with curtains, shutters and blinds, all of which have their drawbacks (notably the nuisance of keeping them clean). What if you could get rid of all these things and, instead, make glass instantly opaque so that it turned clear or dark at the flick of a switch? Windows that change colour with a zap of electricity do actually exist. They're called **electrochromic** ('chromic' meaning colour changing), and they work just like the rechargeable batteries in laptops and mobile phones.

In a rechargeable battery, there are two terminals (one positive, one negative) and a chemical separator called an electrolyte in between. When the battery's charging up, lithium ions (atoms of lithium minus electrons) shuffle through the electrolyte in one direction, storing energy. When you unplug your charger and use the battery to power your laptop, the ions scuttle back in the opposite direction, releasing the stored energy in the form of electricity.

Exactly the same thing happens in electrochromic windows, which are like ultra-thin laptop batteries (made of five separate layers) stuck to sheets of glass. The positive and negative terminals are the outer, 'bread' layers. The middle three layers are a top one rich in lithium ions, a middle one (the electrolyte) that conducts lithium ions, and a bottom one made of crystalline tungsten oxide that receives lithium ions. When the glass is clear the lithium ions sit in the top layer, letting light pass through normally. When you want the glass to turn opaque you flick a switch so the ions move down into the tungsten oxide layer and lock into its crystal structure. That blocks light from passing through, turning the window dark. Reversing the current makes the ions move back, turning the glass light again[105].

A BLACK-AND-WHITE CASE

When I was a teenager one of the coolest things you could own was a pair of **photochromic** sunglasses that automatically darkened in sunlight. TV advertisements are much like gravity – hard to defy, though not impossible – and, when I gave in to temptation and bought a pair with my pocket money, I was fairly unimpressed: they darkened quickly but took forever to lighten again. Worse, they didn't work properly indoors (through car windows) and, for some reason, turned very dark indeed on cold days. Fortunately, science is rarely so disappointing, and I'm still impressed by how these things work, even if their practical performance falls rather short.

You probably know that old-style photographic film (the kind that comes in little light-tight, black containers and has to be 'developed') works using silver halides (simple silver-based chemicals) trapped in plastic. When light hits the film the silver halides magically transform into specs of silver, making dark patches where there's a lot of light. That's why a photographic negative is reversed, with light areas shown dark and vice versa. Toss your developed 'negative' into some bubbling chemicals, beam light through it onto paper and you have an old-style photo made using technology dating back to the days of William Henry Fox Talbot in the 19th century.

▷

Who invented photochromic glass?

Photochromic sunglasses work in a similar way. The original ones used *actual* glass (rather than plastic), and were patented by William Armistead and S. Donald Stookey of Corning Glass in 1962[106]. Like photographic film, they contain a smattering of silver halide crystals (about a tenth of 1 per cent) mixed in with the glass, which darken when sunlight hits them and lighten when darkness falls. How? The ultraviolet light in sunlight powers a chemical reaction that converts the transparent silver halide crystals into tiny specs of opaque pure silver, turning the glass dark (but not opaque) in a minute or two. Nip indoors where there's no ultraviolet and this reversible chemical reaction runs back the other way, packing those silver specs off to the silver halide crystals from which they came – and lightening your dark lenses once again.

Modern photochromic lenses (sold under brand names like Transitions) don't use silver or glass. They're based on complex plastics called naphthopyrans, which change their structure, reversibly, when ultraviolet light hits them. Indoors, they take on a form that absorbs relatively little light. Outdoors, it's a different story: ultraviolet light snaps them into a modified form that absorbs much more visible light, effectively darkening your lenses. Lots of molecules all absorbing light at once, on the atomic scale, block out light in much the same way as blinds closing in front of your window. Take away the ultraviolet light and the molecules soon snap back to their original forms, 'opening the blinds' in front of your eyes.

Glass or plastic, the problem with photochromic lenses is that they're powered by ultraviolet light. Although there's plenty of that outdoors on sunny days, little or none penetrates through ordinary glass, so there's none to speak of inside a car or a train carriage. That's why photochromic lenses don't darken effectively indoors – and why they're much less useful for driving.

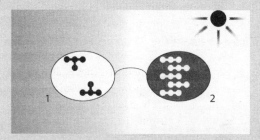

▲ **Photochromic lenses.** *The naphthopyran molecules in plastic photochromic lenses have a particular molecular structure and let most light stream straight through (1). When ultraviolet light (from strong sunlight) hits them (2), they change to a different molecular structure, absorbing more light and effectively darkening the lenses.*

CHAPTER NINE

Saggy Sofas, Squeaky Floors

In this chapter, we discover

> *What connects droopy underwear and anti-wrinkle cream.*
> *Why you can sometimes bounce a glass off the floor without it breaking.*
> *Why shiny shoes last longer.*
> *How materials science can help you share chocolate more easily.*

When Janet Jackson 'boobed' at the Super Bowl in 2004, she made worldwide headlines, and the words 'wardrobe malfunction' burst rudely into the language. You can always count on materials to let you down. How many men scrape daily at their faces with maddeningly ineffective razor blades, wondering why wedges of *stainless steel* that merely have to slice through hair can possibly get blunt at all? How many floorboards have squeaked, creaked or groaned somewhere in the world since you started reading this sentence? Maybe you yourself are shuffling in your seat, right at this moment, wondering why your once-comfortable sofa is sloppier and saggier than it was when new? All these things are signs that no material is perfect – and not even durable things like metal last forever.

It's worse than that, of course. Materials often fail us at the worst possible time. A sudden punctured car tyre or brake failure isn't automatically fatal, but when a chunk of aeroplane skin spontaneously fractures at 10,000 m (33,000 ft) the odds aren't in your favour. In the early 1950s seven catastrophic crashes involving de Havilland Comet aeroplanes (the first commercial jet airliners) were ultimately traced to a design flaw that concentrated stress around the pressurised cabin windows. It's easy to understand why a piece of paper – a flimsy tissue of dried wood fibres – can't withstand the (comparatively) huge force you inflict when your hands rip it in two. It's harder to fathom why durable materials like steel suddenly crack for no apparent reason, or why those beloved Homer Simpson boxer shorts gradually sag and droop after years of dependable use.

We've already discovered how household materials we take for granted – things like wood, glass and glue – succeed so well. But what makes them fail so badly?

Sag

I often think I might jump from a skyscraper or a suspension bridge. Wobbling on the windy roof, dithering by the railings, why not simply leap into the unknown? As a scientist I have complete faith in the theory of **elasticity** and, if I ever did take the plunge, you can be sure I'd have a bungee cord tied tight around my waist. Elasticity is the material-world equivalent of turning back time: the future is just like the past and, within reason, there's no point of no return.

How did 'elastic' ever come to be a material? Lurking in your house you probably have elastic bands and knicker elastic, elasticated sticking plasters and watch straps. In reality, however, there's no such thing as 'elastic': what we really mean is stretchy materials – and almost everything satisfies that description to a greater or lesser degree. In Chapter 1 we noted how the world's tallest buildings wobble up to a metre in the wind, so even something as improbable as a skyscraper is impressively stretchy (if it weren't, it would snap). Properly elastic materials are usually **elastomers**, and rubber, a natural elastomer that powers bungee cords, among other things, is the one we know best. Elastomers are

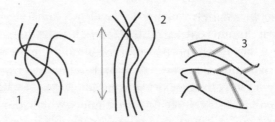

▲ **Why does rubber stretch?** *Latex (raw rubber) has long molecules tangled up together (1). When you pull latex (2) the molecules straighten and untangle (giving plasticity), but also pull back together as soon as you let go (giving elasticity). Latex is easy to reshape with relatively little force and energy, which is why it doesn't last long, and why it turns soft and sticky at relatively low temperatures. Hard, black, vulcanised rubber is latex 'cooked' with sulphur (3) – a process accidentally discovered by US inventor Charles Goodyear in 1839, after many years of fruitless experiments, when he dropped some latex on a hot stove. The sulphur helps to forge tough cross links (grey) between the rubber molecules (black). You need much more force and energy to pull these reinforced molecules apart, so vulcanised rubber is harder, stiffer and more durable than raw rubber. It keeps its strength over a very wide range of temperatures, from about -60 to 200°C (-70 to 400°F), which is why it's suitable for car tyres.*

materials made from large, tangled molecules: when you stretch them, the molecules tease apart and straighten out, but instantly spring back and tangle up again as soon as you let go. Rubber finds its way into some unlikely places. Early chewing gums and bubblegum were chewy because they contained chicle, a natural form of latex rubber, not so very different from what you'd find in a party balloon or a car tyre[107]. Modern gums typically achieve the same chewy effect using synthetic rubbers such as styrene butadiene (probably also in the soles of your shoes) or polyvinyl acetate (found in slippery white glue)[108]. That's why swallowing gum is such a horrible idea.

Plastic or elastic?

While it's true to say that 'elastic' is another name for 'elastic material', it's much more helpful to say that elastic materials (let's call them elastics from now on to avoid confusion) are elastic – because, in science, elastic has a very precise meaning: it's *almost* perfectly reversible stretchiness. When you pull a rubber band and release it, it seems to return to exactly its original size (length) and shape. If you don't get bored easily, you can do this hundreds of times with the same effect.

Just as 'elastic' is a bit of misnomer, so is plastic. We think of plastic as the colourful stuff used to make washing-up bowls and toothbrushes whereas, in strict scientific parlance, the word **plastic** means something more precise: flexible or changeable. Plastic materials (let's call them plastics, for short) aren't necessarily plastic when they're in your hands, but the raw material, usually a bunch of hydrocarbons made from petroleum, was hot and plastic originally. This explains how it could be squirted (injection moulded), squeezed (extruded) or rolled (calendered) to produce the brightly coloured objects sitting in your home. Another reason why plastics are so flexible is that we can manufacture them into extremely fine fibres. The fine nylon bristles in your toothbrush were made by squirting molten plastic through thousands of tiny holes in a shower head-like device called a spinneret. The fibres in nylon stockings are considerably finer. The denier measurement of stockings sounds incredibly impressive when you realise that it's actually the weight in grams of 9 km (6 miles) of the fibre from which the textiles were made (15-denier sheer stockings are made from a thread so fine that 9 km weighs just 15 g/½ oz).

Even when they're finished and in your hands, plastics might still be plastic – even elastic. You can bend a very thick toothbrush handle a tiny bit and it'll revert to its previous shape quite easily. Push too far, however, and you'll see why plastic gets its name. Take a red plastic toothbrush and bend the handle just a little bit too much and you'll deform the handle in an especially ugly way, so that pinky-white strain lines show through. If you look at a strained piece of transparent plastic under a polarising microscope (one that restricts light so that it vibrates in just one direction), you can see amazing rainbow patterns caused by a phenomenon called **photoelasticity**[109]. You can keep on deforming a plastic material until you push it to the point where it fails completely and snaps.

You might also notice an unpleasant smell when you twist or break plastic. That's caused by the strain and the heat releasing gases from the plastic polymers inside. Plastics can give off all kinds of weird smells. Vintage plastic dolls often stink of vomit, for example, because their ancient urea formaldehyde plastic has started to break down. One trick of the trade canny jewellers use to identify unknown plastics is to plunge them in hot water or rub them hard (to heat them), then sniff the gases they give off. Celluloid smells like mothballs; Bakelite and Galalith whiff of formaldehyde or burned milk; and cellulose acetate smells a little vinegary.

Over the limit

Plastics can be elastic, which is usually no problem at all. However, the difficulty with elastics is that they have a nasty habit of turning plastic – either suddenly or gradually. If you stretch a rubbery material far enough, you pull it past its elastic limit. It keeps on stretching but no longer reverts to its original shape.

Interestingly, even metals are elastic: if that weren't the case, driving your car backwards and forwards, jolting it along the road, might eventually cause permanent deformation of the bodywork, the engine and all those nuts and bolts packed inside. Even with rubber tyres and elastic metal springs soaking up much of the energy, vibration is still a problem – although not a fatal one[110]. Washing machines don't shake themselves to pieces because their mostly metal components flex microscopically, absorbing force elastically. If you bang a tuning fork on a table it sings out in middle C because the metal prongs are vibrating elastically, back and forth, at their so-called 'resonant' frequency, several hundred times every second. But you can't push any metal very far without causing what's called **plastic deformation** (permanent change of shape).

Steel is about 200,000 times less stretchy than rubber, which you can easily push to several times its normal length[111]. Rubber is, however, a mere pretender when it comes to serious stretching. Today, the most elastic materials known are **hydrogels**; in a class of their own, you can stretch them to around 20 times their original length[112]. If Dizzy Gillespie's trumpeting cheeks had been made of hydrogels, he could have puffed each one out to the size of his stomach, at least. Even that may be quite a conservative performance. In the 1970s, in *Structures*, one of the best books about materials science ever published, Professor James Gordon produced a table of stiff and stretchy materials, at the top of which came the soft cuticle of the pregnant locust, estimated to be roughly 35 times more stretchy than rubber[113].

Elastics snap when you push them beyond their limit, but they can also fail gradually. A rubber band loses its elasticity over time: stretchiness eventually turns to sag, even if it does take a while. Try wearing a rubber band around your wrist for 2–3 months and you'll see what I mean. To start with it'll fit snugly and tightly, then it'll gradually loosen, sag and stretch rather limply, before eventually breaking apart altogether. Why? Each time around the stretching–relaxing loop, you pull the molecules apart (using a certain amount of energy) without them returning precisely to the position they were in before (or giving the same energy back again). You can certainly feel a rubber band getting hotter if you rapidly stretch it a few times and put it to your lips. The heat given off is wasted energy you can never get back. For vulcanised rubber (the hard, black stuff used in such things as car tyres, originally made by cooking latex with sulphur) it takes many more cycles, and much more force and energy, to cause permanent deformation than it does with the kind of softer rubbers used in balloons and elastic bands[114].

After a few hundred cycles (in the case of an elastic band), the molecules are no longer near enough to pull properly back together. The material still stretches, but no longer shows the same willingness to revert to its original shape. The same goes for all other stretchy materials, from the elastic in a brassiere to the springs in a bed. It even applies to the skin on your face. The multibillion-pound market for anti-wrinkle creams is all based around a simple, inescapable scientific reality: elastic materials, including human skin, always eventually lose their stretch. Between the ages of 10 and 70 your skin loses about a third of its elasticity, mostly by the age of 40. Although mechanical failure (repeated creasing) from laughing and smiling is one reason for this, sun damage is also a major factor. The skin on your exposed cheeks loses its elasticity about twice as fast as that on covered parts of your body such as your upper arms[115].

Snap

Not everything is as reasonable and obliging as elastic or plastic. When I was at school a favourite trick children had was trying to bend plastic rulers until they snapped, suddenly and very shockingly. In those days wooden rulers had largely been replaced by brittle plastic ones that shattered into very sharp fragments, much like glass. This proved such an issue that manufacturers soon switched to 'safety rulers' made from other, more *plastic* and *elastic* acrylics, which bend and strain, and give at least some warning of their impending failure.

Materials that snap are elastic, but to such a slight degree that we barely notice. In other words, their elastic limit is much smaller than that of something like rubber. Even glass is elastic – surprisingly, about twice as much as steel[116]. When you kick a football at a window and it bounces off, the glass does flex very slightly. You can sometimes see this happening when the reflection in the glass wobbles a little bit as the ball ricochets back. It's safest to demonstrate the elasticity of glass by pinging a wine glass with your finger, or moistening your finger and running it around the rim of a glass to make it sing. The glass vibrates a tiny little bit, which is what makes the noise – and it vibrates because it's elastic.

The room for manoeuvre with glass is much smaller than it is with truly elastic or plastic materials, and also with even stiffer and stronger materials such as steel. It might seem paradoxical that glass should be more elastic than steel and yet more breakable, but there's no contradiction. Glass shatters because it takes only a tiny bit of energy to send fatal cracks hurtling through its structure. However, that doesn't necessarily mean it will always break. You can drop a glass or a cup and, if you're lucky, have it bounce away unscathed. How? Watch closely the next time this happens and you'll see that whatever you drop radically shifts position as it bounces. Most of the energy is absorbed in changing the object's motion or spin, so there's less energy for it to absorb internally and less chance that it'll shatter.

There's always a point at which plastic materials give up and snap. You can bend a metal ruler and flex it straight back again (which means that metal is *plastic* – you can deform it). You can do this quite a few times, but eventually the thing will simply snap in two. Paper clips demonstrate this very nicely. You can bend one back and forth maybe half a dozen times or more before it fractures. Why does it break? Because an atomic-scale failure develops in the crystalline structure of the metal. Repeated stress makes the failure spread through more layers of atoms, widening into a bigger crack. Eventually the crack builds to such a size that it's more advantageous for it to spread than not to. At this critical point (technically called the Griffith length), the crack can gain enough energy from the inside of the material to

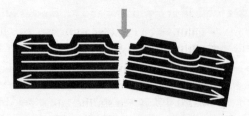

▲ **Using cracks in materials.** *Materials usually fail when tiny cracks gain enough energy to turn into fatal fractures. Chocolate makers build this simple science into the design of their bars by moulding ridges or grooves in the surface where you're supposed to break them. As you flex the two ends, stress builds up along the bar in parallel lines. At the top of the bar the stress is diverted around the grooves and becomes more concentrated there. As a result the bar is much more likely to fracture along the grooves – exactly where you want it to – than in the other places*[117].

propel itself right through, causing a complete and catastrophic break.

Some materials fail because they're weak to start with, others because they're badly designed so the stresses they experience are concentrated at points of weakness. In the case of the fateful de Havilland Comet, the problem was that the cut-out windows and escape hatches created points of weakness in the metal structure of the pressurised cabin. Repeated stresses from the forces of flight caused cracks to widen and spread out from rivet and bolt holes – with terrible consequences. Although de Havilland had tested the plane and figured it good for a 10-year life in the sky, they got their sums wrong. The real stresses and strains were three times greater than they'd found in the laboratory, causing an expectation of catastrophic failure in the planes within a year or two of their taking off[118]. In a crude sense, each doomed plane behaved like a giant paper clip being bent back and forth by the buffeting rush of the wind. It was a good example of what we mean by **metal fatigue**: a sudden failure caused by the cumulative effects of repeated back-and-forth stress.

Squeak

Faint signs of failure aren't always reliable. If you live in an old property with squeaky floorboards you might get so used to it that you find it rather charming. You'll know each board by its own little moan, and you'll learn to tiptoe between the boards when you shuffle around in the night. If you're used to living in a modern apartment building and you suddenly set foot in someone else's creaky old home, it might give you pause for thought: is this place really safe? Does the squeaking signify a fundamental, structural weakness? Is the rickety structure going to tumble down around your ears?

Groaning, creaking wood is usually a healthy sign. It shows that timbers are flexing back and forth, happily absorbing the stresses and strains we put on them. Trees groan when they bend in the wind; this doesn't mean that they're going to fall down (although it might be an early warning of exactly that). Why does wood grumble and groan when other materials don't? With the exception of bamboo, it's the only natural, plant-based material we use to make large, load-bearing structures. It's made of hollow fibres packed inside a dense structure, and as it flexes back and forth these slide and grind against one another, making the groaning noise that you hear. Floorboards also groan as the planks rub past one another and at the two ends where they're nailed down. If wood is particularly stiff, it doesn't strain much when you apply a force, so there's little movement or creaking; the more flexible it is, the more movement between the fibres and the more noise you hear. Changes in temperature and humidity also make wood expand and contract, so creaking and groaning sounds in a wooden home vary from season to season. Floorboards are more likely to squeak in summer, when they're dry and contracted, than in winter, when they're colder, moister and the fibres are more swollen.

SHIP TO SHORE

Old wooden ships used to groan terribly as they pitched and tossed on the waves, but this was a sure sign that they could bend and flex, absorbing the thumping strains of the sea. Ironically, when stiffer, iron-hulled ships were first introduced in the 19th century, they didn't creak in the same way. Terrified engineers feared they'd prove less flexible and more risky as a result. It's easy to see why they fretted if you think about jumping off a wall. Bend your knees and flex your body, and you can absorb the impact very easily; keep your legs rigid and the sudden shock could smash your back. A wooden ship flexing with the waves is a bit like a skateboarder bending her knees to soak up the humps and bumps of the road.

Did iron ships break apart in the sea? The fear ultimately proved unfounded – metal-hulled ships clearly now rule the waves – but there were certainly teething problems. According to maritime author George Goldsmith-Carter, stiffer iron ships 'were more accident prone'. Heavily loaded in rough seas, they jerked violently back and forth, putting too much strain on their tall masts and rigging, and causing them to snap[119]. Interestingly, flexible wooden ships often proved more suited to the freezing conditions in the Arctic Ocean. The greater 'give' of wooden-hulled ships made them better able to break free from the ice[120]. Polar explorer Fritjof Nansen's famous ship, *Fram*, was deliberately built from extremely strong greenheart wood to enable it to survive being frozen in the ice, and lifted up and swept along by it[121].

Wear and fade

When we say something's worn out, all kinds of things could be wrong. A pair of shoes might wear out because you've gradually scraped the soles into holes through friction. Every time you skid or slip you wear away another few layers of atoms, and eventually all that's left is nothing at all. Flexing the leather uppers back and forth is much the same as bending a paper clip, although thanks to the collagen fibres (proteins) inside it, leather is more flexible (plastic) and can go through thousands of back-and-forth cycles before it becomes so weak that it breaks. Raw hide is obviously flexible when it's still attached to an animal, but leather is much more *durable* because the tanning process removes water and causes cross-links to form between the protein molecules inside it, binding them tightly and strongly – a bit like the cross-links we saw in vulcanised rubber[122]. One reason we polish shoes is to lubricate the collagen fibres so that they remain flexible and supple (keeping out water and making them shine for the sake of appearances are other obvious benefits). We'll return to the science of shoes in Chapter 18.

Fade out

When something like a pair of leather shoes wears out, this indicates that we've pushed materials to the point of breakdown – and beyond. Not that it necessarily takes so drastic a failure – a hole or a tear – to make you part company with a favourite dress or shirt. It's just as likely that something old and faded will meet your displeasure too. Why do textiles fade? The usual reason is that they're dyed – and dyes are essentially *chemicals* that become impregnated in their fibres. Sunlight fades things primarily because it contains ultraviolet light (the same high-energy, high-frequency light that gives us sunburn). When ultraviolet light hits dye molecules, it causes **photodegradation**: the molecules rearrange into different forms that don't reflect light in the same way. You might have noticed that supermarket receipts (and faxes) quickly fade in the Sun. That's because they're not printed with ink, but using heat-sensitive chemicals called **leucodyes**, which have two forms: leuco (colourless) and non-leuco (coloured). When a thermal printer runs its heated printhead

over heat-sensitive paper, words appear, because the leucodye
is converted from its colourless variety into its black, non-
leuco form. Ultraviolet light makes the leucodye molecules
change back again, effectively bleaching out the colour and
making them fade[123].

Plastics also photodegrade, often quickly and drastically.
When London's Wembley Stadium was expensively
refurbished in 2007, around 90,000 bright-red plastic seats
were installed, but 17,000 of them very quickly faded to a
washed-out pink through photodegradation[124]. You've
probably noticed that clear plastics often go cloudy with
time or sometimes turn a horrible yellow. Again,
photodegradation is to blame. A transparent plastic (such as
PET, used in water bottles) allows light waves to wiggle
through largely unchanged. When plastic turns yellow the
molecules inside have been transformed so they transmit
only part of the light hitting them and absorb the rest. The
light they transmit is a mixture of red and green, giving old
plastic its familiar yellowed appearance. A side effect is that
the plastic turns brittle and is far more likely to crack.
Although that can be annoying, it's a useful way of getting
plastic to break apart in the environment; without the help
of natural effects such as photodegradation, plastics would
stick around virtually forever.

Rust and rot

Not all material failures are physical: some are chemical or
biological. When your car starts to rust and fall to bits, it's
failing *chemically* because iron exposed by scratched or
damaged paintwork has reacted with water and oxygen in
the air to form iron oxide, better known as rust. Unlike iron,
which is strong and sturdy, iron oxide is a flaky powder, so
areas of rust form major points of weakness in bigger pieces
of iron or steel. As we saw in Chapter 7, adding chromium
to iron to make stainless steel forms a protective oxide 'skin'
around the iron atoms that stops oxygen and water from

attacking them, keeping them relatively (although not entirely) rust free. Paint does exactly the same job. When I was younger I used to think people painted cars to make them look attractive, but that benefit is really only incidental: the main reason is to keep out damp and rust. Cars and ships made from aluminium don't need painting at all: an oxide layer naturally forms around them, keeping them shiny and safe.

Rot is analogous to rust – a biological bugbear that nibbles away at wood, rather than metal. The treatment is exactly the same: create a barrier between the wood inside and the fungus-filled air outside, and you can keep rot at bay pretty much indefinitely. Paint the wooden window frames of your home and you'll soon have a different problem, however. Instead of worrying about the material failure of the wood itself, you have to switch your attention to the failing paint as it 'breathes' in and out (expands and contracts) in hot and cold weather, before it cracks, breaks and flakes away, exposing the unprotected wood beneath to more serious damage. Creosote (the kind made from coal tar) contains more than 100 chemical compounds and protects wood from rotting by an entirely different method. Instead of creating an external barrier, it works because it's naturally toxic to the micro-organisms that cause rotting (stopping fungi, insects, mites and spores from attacking things like fence posts and telegraph poles). The trouble is that creosote leaches out from wood into the environment – and it's both toxic and carcinogenic[125]. So by solving one problem, you create a different one entirely.

Trees can swing and sway for hundreds of years where wooden window frames barely last a few decades. Why does natural wood (inside a tree) rot so much more slowly than harvested timber used in the doors and windows of your home? Tree bark is usually waterproof and rich in oils that bust bugs like a natural creosote. That's why, in Lapland, birch bark is traditionally used to make waterproof clothes and shoes[126]. Inside a tree the cellulose (the basic plant material) and lignin (a complex polymer in the walls of plant cells) form a very dense material that's superbly resistant to damp penetration and the microbes it powers. Unprotected timber in buildings is more likely to be warmer and wetter – just right for a fungal feast.

PLASTIC HEAL THYSELF

What's more annoying than someone scratching your car in the street? Wouldn't it be great if cars had the sense to realise this had happened, drove themselves to the repair shop and got resprayed automatically before you came back? That's not likely to happen – but the next best thing, paint that repairs itself, is already here. Many modern paints are essentially plastics (acrylic polymers containing coloured pigments), which you spread in a thin layer over some relatively vulnerable material such as iron or wood. Materials scientists have already developed a range of plastics with built-in repair mechanisms that automatically respond to minor damage: they're called **self-healing materials**.

How do these materials work? Inside the plastic there are **microcapsules** of glue and a **catalyst** (a substance that speeds up a chemical reaction). When the plastic gets scratched or cracked, the capsules burst open and, with the help of the catalyst, leak out the glue (or some other kind of healing agent) that promptly seals up the damage. Other self-healing plastics have repair tubes inside them, a bit like blood vessels, which are linked to a pressurised reservoir filled with glue (or repair gel). If the material cracks, it opens up some of the repair tubes, releasing the pressure so that the reservoir pumps the magic chemical up to where it's needed.

It's easy to see how minor cracks and scratches can be fixed this way, but what about more substantial damage? The US space agency NASA has been developing large-scale self-healing materials that can automatically repair bullet holes in fighter planes and meteorite impacts in spacecraft. When an incoming bullet hits the plastic fuselage, its enormous energy heats up the material just enough so that it briefly becomes a flowing liquid in that very specific area. As the bullet passes through the plastic bends and flows around it, then snaps back into position and seals up again, instantly repairing the damage. In NASA's lab tests self-healing materials have proved capable of resealing themselves when things strike them at micrometeorite speeds of up to 18,000 km/h (11,000 mph) – or about 20 times the cruising speed of a Jumbo Jet[127].

▲ **How self-healing materials work.** *A plastic contains embedded capsules of glue (grey circles inside large black circles) and a catalyst (white dots), which speeds up its operation. When a crack spreads through the plastic, it eventually ruptures one of the capsules, releasing the glue, which flows down the crack and seals it up.*

CHAPTER TEN

Light Delights

In this chapter, we discover

> *Why you can make light from a grain of sand, but not from a Scotsman's beard.*
>
> *What makes a candle more dangerous than a volcano – and how many fireflies you'd need to take its place.*
>
> *How many particles of light a torch pumps out in each second.*
>
> *Why you can see your face in shiny shoes.*

Ever get the feeling that you're not alone? Look up from this page for a moment and glance around the room in which you're sitting. What do you see? Sofas, computers, books, toys, rubber plants, wine glasses, the odd child scattered here and there – everything – and yet, maddeningly, *nothing*. Because what you see is ultimately and *only* light. Curiously, although light is all you can ever see, you can't see light at all. What does it look like? Where is it? Keep looking further, longer and you'll see something completely different: nothing whatsoever. That's what space looks like in the absence of light. You could say it's a big fat lump of emptiness, but that would be something of a contradiction. Space is the darkest thing you can imagine.

Light has always been a riddle, as flighty and fleeting as a butterfly – although some people were confident enough to think they'd pinned it down. Isaac Newton, the brilliant but petulant English scientist we met in Chapter 1, had already figured out much of what we know about light by 1704[128]. That's when he published a daringly comprehensive study of the subject called *Opticks*. Most of Newton's work holds true to this day, including his dazzling discovery that you can split 'white' light into a rainbow of colours, and his gut feeling that light is made up of microscopic cannonballs ('corpuscles') of energy firing into our eyes[129]. Newton's contemporaries, notably Robert Hooke and Christiaan Huygens, argued that light was much better explained as a wave motion, a curiously invisible rippling through empty space, too fast and tiny for us to see *in itself* but obliging enough, nevertheless, to reveal the magic of everything else. To this day scientists (and, as you'll see in this chapter, science-book writers) flit between describing light as a stream of particles and as a train of waves[130].

Today we call the particles **photons** and use them to explain all kinds of household things, from the magic eye in the burglar alarm, to the solar panel on the roof. The way we drone on about photons, you'd think we'd cracked light's riddles long ago, but it takes true wisdom to admit doubt. In the early 20th century, when many people thought there was little more for physics to achieve than tying up the odd loose end here and there[131], Albert Einstein spotted some gaping holes in our knowledge. His radical new theory, **relativity**, guaranteed that there'd be plenty of work for academics for generations to come. (One of its central ideas – that the speed of light is dependably constant, however you measure it – leads to all kinds of puzzling ramifications, including rockets that put on weight as they go faster and identical twins who age at different speeds.) Einstein was confident enough to make no secret of his ignorance: 'Every physicist thinks he knows what a photon is. I spent my life trying to find out what a photon is and I still don't know it.'[132]

Over a century after Einstein's crazy theory made its baffling debut, we sit in our cosy homes, bathed in light, bombarded by photons, with little idea of what they are or why. There's a huge amount we don't know about light, but much that we do.

What is light?

The short answer is 'energy we can see' – although, like many four-word answers to three-word questions, that's bound to be a gloss. Most of the problems we have getting to grips with light stem from our tendency to assume that it's utterly remarkable and probably unique. And it certainly is, for two very different reasons.

For most of us light is our main source of information. A third to a half of the brain's crinkly cortex passes its time processing the data our eyes hoover up from the world[133]. Scientists find light hugely important for a quite different reason. Ever since Einstein they've known that there's something special about the speed of light – the fastest anything can ever go. Interestingly, the speed of light pops up in scientific equations that have nothing to do with seeing or sensing the world, most famously Einstein's own equation $E=mc^2$. This tells us that energy and mass are the same thing, somehow connected by light – which is an astonishing idea that still baffles most people beyond belief[134]. How can, for example, a beer belly (mass) be the same as the song of a spring-morning blackbird (energy). To constrained, conventional minds, it makes no sense. The explanation? It's far too facile to think of light as we do, as the output of a giant cosmic torch, racing around space and the world, lighting the darkness purely for our convenience. Light exists for some much more fundamental reason – and our vision (an accidental advantage gained through evolution) helpfully took advantage of it.

In other words the word *light* means something quite different to scientists and to ordinary people. This is a much more philosophical way of looking at things – of shedding *more light* on light, so to speak. Humans have an egocentric view of the world. Peeking past the curtains of our eyelashes, squinting through squidgy little windows at the light pouring in, we assume that the world we see is all that exists and construct the meanings of our lives on a human scale.

Scientists see bigger and smaller realities – astronomical distances measured in light years (9.5 trillion km/6 trillion miles, the distance light can travel in a year); and nanoworlds of bonded atoms and molecules, where human ideas of meaning don't merely vanish, but could never possibly exist[135]. For scientists, light isn't simply 'energy we can see'; it's more like 'energy of a very particular kind that travels at a precise and incredibly fast rate, a tiny proportion of which our eyes just happen to be able to detect'.

The light you can't see

We got a strong hint of how skewed our world view is in the previous chapter when we encountered infrared and ultraviolet, the two kinds of light that 'bookend' human vision. Infrared is just too red for our eyes to detect; ultraviolet is just too blue. But what happens if you keep tweaking light waves? What if you take infrared and make it even redder? Take some light waves (which have a wavelength of about 550 nanometres, roughly 100 times thinner than a human hair), and stretch them hundreds of thousands of times. What you get are microwaves that can cook food or, less obviously, zap mobile-phone calls invisibly back and forth to your friends. Stretch them as much again and you get radio waves, exactly like the ones that bounce TV and radio programmes into your home. Now go in the opposite direction. Take some ultraviolet light and try making it bluer. Take those waves and squash them, as hard as you can, in some kind of microscopic vice. Squeeze them down to a thousandth of the size and you'll get X-rays. Keep going and you'll make gamma rays, which are effectively super high-energy X-rays[136].

You might reasonably conclude that all these different names – light, infrared, ultraviolet, microwaves, radio waves, X-rays and gamma rays – suggest that we're talking about different things: that's what names are for. Yet the difference is simply one of degree. Gamma rays differ from X-rays and microwaves in the same way that red light differs from green – a mere difference in the size of the waves and the energy that they carry. Together, these things make up what we call the **electromagnetic spectrum**, which has light (visible light – the kind our eyes can see) perched right at the centre.

The names we give to different parts of the spectrum are arbitrary and historic: big or small, it's all the same. Whether we can see it or not, all this stuff is what scientists mean by light. It's all energy, racing around at the same speed, surfing on waves that stretch from the gigantic (the biggest radio waves can be hundreds of metres wide) to the subatomic (like gamma rays, thousands of times tinier than atoms). If our eyes evolved to see X-rays, we might make out shadowy bombs in suitcases or find the cracks in our own broken bones (we'd need to make the X-rays in the first place, of course, but that's just a detail). If we could see microwaves and radio waves, and our brains could decode their digital signals, we wouldn't need television or radio sets to watch *Doctor Who* or listen to *Woman's Hour*. We could see and hear those things right inside our heads.

All light is electric

When we talk about electric light we usually mean the kind that floods the room at the flick of a switch. Thomas Edison liked to believe that he invented this stuff and you might like to think so too, but you'd both be wrong. All light is electric – and it always has been, long before electricity was discovered and way before there was even a word for it.

Whether we're talking about flickering candles and crackling fires or radio waves, microwaves and X-rays, all light (in the very broadest sense of that word) is made by electricity and magnetism skipping through space, hand in hand. If you could slow down light and inspect it on an atomic scale, you'd see that it travels when energy ripples along in undulating patterns – waves – of electricity and magnetism. Picture the space between your eye and the Sun as a giant ocean of electromagnetism. When sunlight zaps into your eyes, what's happening is that electricity and magnetism are rippling their way through that emptiness just like waves roaring over the sea.

Why can't we see light?

Where ocean waves carrying surfers pootle along at about 40 km/h (25 mph), light is a bit more fleet of foot: 27 million times faster, to be exact[137]. It can leap 300,000 km or 186,000 miles (seven times around the Earth) in a second, so it takes a matter of minutes to get here from the Sun. That's one reason why we can't see light rippling through space the way we can watch waves on the ocean. Another reason is that light is pretty tiny stuff. Taking visible light, for example, each wave is typically a few hundred nanometres long (a few thousand times bigger than a typical atom), so the chances of seeing a light wave are, quite literally, *vanishingly small*[138].

What about photons? If light is made of photons, why can't we see them? This is where we enter the surreal, Alice in Wonderland world of **quantum theory** – the weird set of ideas that helps us understand how things behave on the atomic scale. It turns out that photons are vanishingly small too and, indeed, have no mass: they are pure energy. It's relatively easy to calculate how much energy a single photon carries (although it varies according to the colour of the light). A helium-neon laser, for example, fires out a steady stream of red photons, each with an energy of about 0.000000000000000003 joules[139]. An ordinary torch (flash-light) bulb pumps out something like 2 million trillion (2,000,000,000,000,000,000) photons each second[140]. How can you visualise such a massive number? Imagine if every person on Earth were made of 300 million tiny 'people particles'. Add up the total number of these particles on the planet and that's equivalent to the number of photons you'd get from your torch bulb every second. Bearing in mind how wide the beam is, you get some idea how big photons would be if they really did have a measurable size: 300 million times more photons crammed into that gleaming cone, every single second, than there are people on Earth.

What makes your home so light and bright?

In Chapter 2 we discovered that energy doesn't appear out of nowhere or vanish into nothing. Since light is a kind of energy, it follows that it too must originate from *something* and the energy that it carries must come from *somewhere*. Whether you make light with a torch, a grubby old power-cut candle you found under the sink or a modern energy-saving fluorescent bulb, it has to get its energy from somewhere.

So where does the light come from? It comes from atoms. As we've already seen, an atom is a nugget of matter mostly packed in its centre, or nucleus, which is stuffed with two kinds of fat particles, protons and neutrons. Around the edge of an atom there are those much more flighty particles, electrons, and usually exactly as many electrons as there are protons. Crudely speaking, we can think of the electrons sitting at certain distances from the nucleus in concentric shells, which are a bit like the layers in an onion. However, all the pictures of atoms we draw (or see) in books are way off the mark. As Brian Cathcart's book about the history of atom splitting vividly puts it, the nucleus of an atom is about as big as a 'fly in a cathedral'; a pea in the centre circle of a soccer stadium is a similar analogy[141].

Forget peas and flies and soccer for a moment – and concentrate on electrons instead, which fill the rest of the space. If you fire energy into an atom, it becomes 'excited' and one of the outer electrons leaps out to a shell that's a bit further from the nucleus. Now it takes some effort to push an electron further out (just as it takes effort when you climb a ladder and move your body a bit further from the Earth), so that's how the atom absorbs energy. In much the same way that people wobbling up ladders feel unsteady and long to return to the ground, so excited atoms are unstable and do their best to return to their 'ground state' as soon as they can. Unfortunately, an excited atom can only get back to its original state by shedding the energy it's gained (think of it as a kind of bank robber desperate to ditch his loot). It does this by spewing out a photon of light roughly a nanosecond (a billionth of a second) or so after it absorbed the original energy, whereupon the electron drops back to its original shell. That, simply speaking, is where light comes from: atoms take in energy (supplied by something like heat or electricity), become unstable and give out light. Almost every kind of light you can think of is made by a variation of this simple process, which is called **spontaneous emission**.

Sunlight
If it's daytime where you are and you're bathing in natural light, you might be interested to know where it came from.

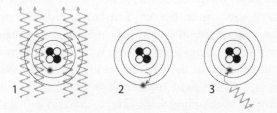

▲ **How atoms make light.** *Suppose you heat an iron bar in a fire until it's glowing red hot. What makes it red? Iron atoms inside the bar soak up the heat energy (wavy lines) from the flames (1). Individual iron atoms absorb the energy by pushing their electrons into higher orbits (2), making the atoms 'excited'. Excited atoms are unstable, so they return to their ground state after about a nanosecond or so. They give out the energy they took in (originally in the form of heat) as photons − light particles − we can see (3). In the case of hot iron the photons carry red light, which is why the bar looks red hot.*

About 8½ minutes ago it was up in the Sun, 150 million km (93 million miles) away. What we think of as sunlight is a highly successful export product of a nuclear factory deep in space. For billions of years the Sun has seethed and boiled with nuclear reactions that fuse the most basic of atoms (hydrogen) into slightly more complex ones (helium), unlocking energy in the process. The energy produced by this nuclear fusion excites atoms and spews out light, including ultraviolet that burns our skin and the visible sunlight that paints the smile on a friend's face. Chew on that thought for a moment: once upon a time, the words you're looking at now were, effectively, a couple of atoms copulating in the Sun[142].

Candlelight

Before electricity came along people had to get light from fire. All light, in other words, was **incandescent**: if you wanted light there was no choice but to make some heat as well. When you burn wood, flare up a match or set fire to a tricky candle wick, you're kick-starting a chemical reaction (combustion) in which the fuel (wood, wax, coal or whatever else you're burning) is converted into simpler atoms with a release of energy – heat and light. Some of the energy excites atoms in the fuel. When they calm down again, they beam off their energy as infrared (light that feels warm) and visible light (the red, orange, yellow or white-hot glow of something that's blazing hot). When it comes to breathless inefficiency, candles are in a league of their own. That feeble flicker might seem barely bright enough to read by, yet it's powered by a flame that can reach temperatures of 1,400°C (2,550°F), which is significantly hotter than volcanic lava[143].

Filament light

Old-fashioned electric lamps (and simple torch bulbs) still use incandescence, but instead of getting their energy from blazing fuel, they soak it up from the electric current zapping through them. When you force electricity through a thin wire, you make the electrons inside march past the atoms that would normally contain them. The thinner the wire, the harder it is for the electric current to flow – and that's what we mean when we talk about electrical **resistance**. Using the right balance between current and resistance, you can make electricity heat up a filament of wire so much that it glows red or white hot. The catch is that you're falling just short of setting fire to the filament, which is why it has to live inside a bloated glass bulb starved of oxygen. Without this cunning innovation, an electric lamp would still produce light, but probably only for a few minutes. With the bulb clamped in place, the electricity buzzing through the filament generates heat, which excites atoms so they give off light.

The long-lasting filament was the bit that Thomas Edison perfected when he finally patented his electric lamp in 1880. The way history tells the story now, you'd think that the greatest gadget of all time – a bright-burning light bulb is the very symbol of inventive genius – was one of many fantastic ideas queued up and waiting for history's all-time favourite inventor to give it the time of day. But that's a myth. Dozens of other people had already tinkered with electric lights before Edison showed up, and even his contribution was more 'perspiration than inspiration'. With a chained-up bear standing guard outside his pioneering research lab at Menlo Park, New Jersey, Edison tested about 6,000 different filament materials, from bamboo and paper to cotton and even red hair from a Scotsman's beard, before finally settling on tungsten metal locked in an air-tight glass bulb – the form in which incandescent lamps survive (just about) to this day. The glass bulb is the real secret, of course: without it, almost any filament is destined to burn bright and die, like a precocious Hollywood star[144].

Fluorescent light

The main problem with incandescent lamps is that they throw off heat as well as light; at least 90 per cent of the electricity they consume is wasted heating up the filament and the air around them. Modern energy-saving lamps are much more efficient because they make just as much light as incandescent lamps with nothing like as much heat. But if there's no heat, where does the energy come from to provoke the atoms into producing light? It comes from atomic collisions.

A fluorescent lamp is a sealed glass bulb filled with gas that has two terminals (**electrodes**). When you switch on the power the gas atoms split up into **ions** (atoms missing electrons) and loose electrons, and scurry about inside the bulb. The atoms, electrons and ions frequently crash into one another. Every time they collide, the energy from the smash causes a bit of atomic excitement, and there's a flash of *invisible*, ultraviolet light. We wouldn't normally be able to see this, of course, but the insides of the glass tubes of fluorescent lamps are dusted with white powdery stuff called phosphor. When ultraviolet light hits the atoms of phosphor, they get excited, jump to higher energy levels, then return to their original ('ground') state. Instead of giving off the ultraviolet light they took in, however, what they give out is visible light. In other words, the white phosphor coating transforms the ultraviolet light into ordinary light that we can see. If you've ever wondered why fluorescent lamps are always white, and never transparent like incandescent bulbs, that's the reason.

Neon lamps work in a broadly similar way, but they're filled with a gas (neon) specifically selected so that it creates red light when the lamps are excited by the electric current that passes through them. Although neon lamps is their generic name, not all of them are filled with neon gas. Other, similar 'noble' gases such as xenon or argon (or even a mixture of different gases) are used to create all kinds of colours and effects.

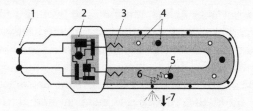

▲ **How an energy-saving lamp works.** *Energy pumps from the power supply (1) into an electronic circuit that boosts the frequency of the current to prevent any noticeable flicker (2). The circuit drives a pair of electrodes (3) that energise the gas in a sealed tube. The incoming energy ionises the gas (4), which means that electrons are knocked from the atoms inside it to make ions (atoms missing electrons, coloured black) and free electrons (coloured white). When the atoms, ions and electrons collide (5), ultraviolet light is emitted (6). As this passes through the white walls of the tube, atoms of phosphor (black dots) convert it into visible light (7).*

You don't need electricity to make fluorescent light. Anything that can fire atoms with energetic enthusiasm can, in principle, re-emerge as a flash of light. That's why certain hard sweets (such as Life Savers) flash in your mouth when you crunch them in the dark. Your teeth supply the energy, and the sweet's flavouring (typically wintergreen oil, methyl salicylate) converts that into visible light – much like the phosphor coating in a fluorescent lamp, but producing blue light instead of white.

LED light
Why are some lights more efficient than others? The Law of Conservation of Energy tells us that the light we get from a lamp comes from the energy we put into it some other way. The more simply we can make electrons produce light, the less energy we waste, and the more efficient the light source will be. That's why fluorescent lights (which just smash atoms together) are so much more efficient than

incandescent ones (which require us to heat up a filament), and why those are more efficient, in turn, than candles (where a giant slab of wax has to melt to vapour before any light appears).

It follows that the most efficient light of all would do nothing more than shuffle electrons around. That's what happens in LED (light-emitting diode) lamps, which are even more efficient and longer lasting than CFLs (compact fluorescent lamps). A diode is the simplest possible microchip and it works like a kind of electronic one-way street. Electricity can flow through it, properly, in only one direction. As the name suggests, a light-emitting diode is a particular kind of diode that gives out light when electricity zaps through. It's the silicon-chip equivalent of an incandescent bulb.

How does it work? You make a diode by shovelling up some sand and extracting its major ingredient, silicon (the black speckles you see if you poke around in a handful of dry sand)[145]. Grow the silicon into a crystal and you have the raw stuff from which we make computers, phones and virtually every other electronic gizmo. Normally, silicon doesn't conduct electricity particularly well so, to make it work in an LED, we have to add impurities. First, split your silicon in two. Add one kind of impurity (boron) to one half, and you'll rob some of your silicon atoms of their electrons, leaving them with 'holes' and making them slightly positively charged. Do the opposite to the other half of your silicon, adding a different impurity (antimony), and you'll get silicon with too many electrons, which is a little negatively charged. The clever bit is when you join the two pieces of impure ('doped') silicon back together. Wire this thing – which is called a junction diode – into a circuit and you'll find that electricity flows through it in only one direction. As it does so, the extra electrons jump across from their side of the junction, recombine with the holes on the other side to make complete atoms and give off a visible sigh of relief: a photon of light. That's pretty

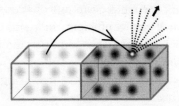

▲ **How an LED lamp works.** *The magic happens at the junction between a lump of silicon (white) rich in electrons (grey dots) and a similar lump rich in 'holes' (black dots), which are the gaps left by missing electrons. Connect a battery across the junction and the electrons leap across the divide, recombining with the holes to give off light (dotted lines). The only moving 'part' in an LED is the leaping electrons, so it's superbly efficient at converting electrical energy into light.*

much all there is to an LED – electrons jumping about inside silicon – and that's why this is such a super-efficient way of producing light: virtually no energy is wasted whatsoever.

Glow-worms and fireflies
Living things make light too, not to read in the dark or see where they're going, but to attract mates or scare off predators. On land glow-worms and fireflies flicker their tails like living flares; in the oceans huge shoals of ghostly glowing squid, sardines and spooky starfish light the dark depths with astonishing electric blue displays. The scientific name for these underwater fireworks is **bioluminescence** and, like every other kind of illumination, the light itself comes from excited atoms shedding excess energy.

Where this type of illumination differs is in where the energy comes from in the first place. Fireflies don't set fire to themselves, like candles, or run off batteries, like torches. Instead, they cause chemicals (luciferin and luciferase) stored inside their bodies to react together and make the light that way. They're a bit like those emergency glow-sticks you see propped up in train carriages. Snap one of those in half and you break some small glass containers inside, mixing together chemicals and making light as a by-product of reactions. You don't get much glow off a worm or fire off a fly; you need about 100 fireflies to make as much light as a single candle[146]. Nevertheless, the trick is still impressive.

THE SCIENCE OF SHINY SHOES

When I was young I could never understand why I was constantly being ordered to polish my shoes. What on Earth was the point? Clean shoes came at the expense of a grotty kitchen floor – and the polish always stank like a chemical plant. Scraping along in the gutter, polished shoes were literally beneath contempt. With my nose whistling through the wind, I never looked down or cared what went on at ground level. These days, squelching through the English countryside, I understand the difference that a regular coating of waxy polish can make to my sturdy walking boots. Not only does it keep the rain out, but it makes the shoes last about twice as long (as we saw in Chapter 9, it lubricates the leather and stops it from cracking as it flexes).

Why do polished shoes shine? Ordinary leather looks dull and shabby because it's covered in microscopic scrapes and scratches. When light rays hit a surface like that, they scatter in all directions. Some of them bounce into your eyes, which is why you can see your shoes – and why they look brown, blue or whatever colour they are. But the light rays don't reflect in a methodical way: stare at your shoes and the rays from your face get scattered and scrambled. It's reflection, Jim, but not as we know it; it's called **diffuse reflection**, which means that the light rays shoot off in all directions. That's very different from what happens when you peer in a flat, polished mirror. Here, every light ray heading out from every one of your pimples and wrinkles hits the glass at a precise angle and bounces back into your eyes at exactly the same angle, so your face is very faithfully reflected. This is **specular reflection**.

When you polish your shoes you're adding a thin coating of wax that fills in the lumps and bumps, and creates a more uniformly reflective surface. Polishing something is the equivalent of filling potholes in a road. Incoming light rays don't bump about as much, so they reflect back more regularly – and the waxy surface behaves more like the glass on a mirror.

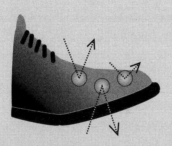

▲ **Why polished shoes shine.** *Shabby shoes scatter light rays at all kinds of angles, preventing a coherent reflected image from forming. Polish those shoes shiny – and they work more like a mirror. Each atom (large circle) in the gleaming wax polish snatches up incoming light energy, becomes excited and unstable, and spews the energy back again about a nanosecond later. Just like in a mirror, light rays are reflected at the same angles as the ones at which they approached. This turns the diffuse reflection of dirty shoes into something much more like the specular reflection you get from a mirror.*

▷

If you're the sort of person who likes shoes you can see your face in, but hates the 10 minutes or more that it takes to achieve that effect, spare a thought for the people who make telescope mirrors. Perkin Elmer, the company that built the mirror for the Hubble Space Telescope, which measured 2.4 m (7 ft) in diameter, initially took about 18 months to do the job. Why so long? Because when you're dealing with something as small as light, and attempting to capture things happening half a Universe away, bumps and scratches the size of atoms make a big difference. In the case of the Hubble mirror, Perkin Elmer tried to polish to an accuracy of 20–30 nanometres (roughly the diameter of 50 atoms stacked on top of one another). If the mirror were scaled to the size of the Earth, bumps like that would be smaller than the palm of your hand (about 10 cm/4 in across). Unfortunately, they were so obsessed with polishing it smooth that they failed to notice they'd made it the wrong shape.

But that's another story entirely[147].

CHAPTER ELEVEN

Radio Gaga

In this chapter, we discover

> *Why radio masts must soar to the sky – but mobile phones can still fit in your pocket.*
> *How a 'worm' helped to invent modern, global communications.*
> *Why you can't cook chicken korma with a smartphone.*
> *How the Victorians almost had mobile phones in 1880.*

Earth, air, fire and water – the classic elements of ancient times. Some early thinkers, shrewd enough to realise that four elements were too few, pondered over the existence of a possible fifth, *quintessence*. Often known as aether (or, even more charmingly, the 'luminiferous aether'), it was a mystical, magical, airy-fairy sort of stuff, a kind of nothing-stuffing that padded out our empty Universe, supposedly making it possible for light to dash from place to place. It wasn't until 1887 that scientists finally realised aether was a myth – and that light could zap from here to there without anything to whisk it on its way[148]. That insight fitted neatly with a growing understanding of how light was really a kind of electromagnetism – a wobbling pattern of electricity and magnetism – and led to the remarkable, early 20th-century development of a practical way to fire information across the planet in the blink of an eye: radio.

It wasn't always thus. Step back a few centuries and no one would have cared less about communicating with people on distant continents. Life was more local and, most of the time, speaking or shouting was good enough to get you heard. Sound is reasonably nippy, bumbling along at about 1,200 km/h (760 mph), but its limitations quickly become obvious. A lightning strike on the other side of town will flash seconds before we hear the rumble of its thunder. The light reaches us instantly, but the sound, which sets off at the same time, takes about three seconds to trudge each and every kilometre (five seconds to go a mile).

If we had to rely on sound alone, modern-day communication would be impossible. Jumbo Jets hurtling overhead seem to inch through the sky – and sound is only about 30 per cent faster. Imagine holding a telephone conversation between New York City and Los Angeles (separated by about 4,000 km/2,500 miles) if your phone merely carried words – the same way as ordinary speech – at the speed of sound. Picture your words flying back and forth through the clouds only slightly faster than jets. It would take more than three hours for each utterance to go from speaker to speaker, and just as long for the reply to go back. Conventional conversations that swap 30 sentences in a couple of minutes would end up stretching over four days and nights. Light, on the other hand, shrinks Earth to a pinpoint. A beam of light can reach the moon in a second, the Sun in less than 10 minutes and Mars in 20 minutes at most. Only when we start thinking about the far reaches of space do light's limitations become apparent.

Look around your home right now and you'll find that pretty much all the useful information beaming into it sails on a magic carpet of electricity and magnetism. Radio, television, telephone and the Internet all rely on electromagnetic waves. Even staring through the window, daydreaming to a backdrop of distant events, relies on your eyes and brain sifting through drifts of incoming light. Is it simply a coincidence that electromagnetism is such an effective information carrier? Or is there something more subtle that explains it? And just what makes it so fast anyway?

Light was always right

It's tempting to think that speed-of-light communication is something radically new, although even mobile phones, which beam data back and forth using radio waves, are more than 40 years old[149]. A bit more thought reminds us that light-speed communication is classic and timeless. How about smoke signals and semaphore flags? And beacons puffing away on hillsides to signal foreign invasions? What about winking lighthouses and those tilting-arm signals on railways? All these things transmit messages visually – at the speed of light – even if they can't send them very far.

Ironically, the development of electricity made communication slower, not faster – at least to begin with. It takes longer to buzz a message down a wire than to ping it through the air as a beam of light, because nothing goes faster than a blistering flash: the speed of light is the quickest thing there is. Even so, it was electricity that revolutionised communication. It took numerous attempts to lay an undersea (submarine) cable between the UK and North America but, once complete, it instantly cut the time to send a transatlantic message from around 12 days to just 16½ hours (for the very first, laboriously transmitted public message signed by Queen Victoria). Later messages were sent in a matter of minutes[150]. That sounds extremely impressive until you stop to think that a beam of light can make the journey from London to New York City in just two-hundredths of a second[151].

Once 19th-century pioneers such as Michael Faraday and Thomas Edison figured out how to generate electricity in copious quantities, it became the obvious and practical way to transmit information, but with one very important drawback. Instant, local communication at the speed of light gave way to slower, long-distance messaging at the speed of dark – the slightly more sluggish trudge of electric information. Telegraphs linked countries and continents, but it still took a surprisingly long time to send messages, not least because words had to be translated to and from Morse code for transmission, and the messages themselves had to be relayed at relatively modest speeds to prevent the electrified dots and dashes they contained from getting jumbled up[152]. Telephones improved things, incrementally, but long-distance calls were still complex. Each conversation had to be routed by human operators yanking cables in and out of switchboards.

Interestingly, telephone pioneer Alexander Graham Bell knew instinctively that his famous invention was an inconvenient way of sending messages over long distances. Spoken words followed wires, so you could only hold conversations between two geographically fixed points determined by the phones at either end. In 1880, four years after his original patent, he sketched out his prototype 'photophone', which was meant to send sounds and pictures not down crinkly wire cables, but flickering through empty space in light waves. It was an inspired nod towards the present-day mobile phone. If he'd pulled it off, buttoned-up Victorians in steam-train carriages would have been able to annoy one another with inane mobile-phone chit-chat, just like people do today.

THE VICTORIAN MOBILE PHONE

Think telephone – think Bell? Although Scottish-born Alexander Graham Bell gained the glory for developing one of the world's most far-reaching inventions, there's long been debate over who got there first. Elisha Gray patented a similar invention on the very same day in 1876, while another rival, Antonio Meucci, had started developing a primitive phone back in 1849 – almost 30 years earlier[153]. Bell has a stronger claim to inventing a more modern version of the telephone many of us now rely on: the mobile phone (also called cellular phone or cellphone), which sends and receives its calls not with cumbersome coils of cable but through invisible, high-energy radio waves.

Recognising the limitations of the wired 'landline' phone, Bell and his assistant Charles Tainter came up with their photophone just four years later. The basic idea was simple:

the speaker's voice was channelled into a large horn, which made a grating vibrate up and down. A beam of light, shining through the grating, flickered on and off and became a kind of encoded version of the speaker's voice, shooting over to a distant receiver unit, which reversed the process and reproduced the original sound. In tests, Bell and Tainter successfully used a light beam to transmit sounds for a distance of more than 200 m (700 ft)[154].

Bell considered the photophone his greatest invention – better even than the (now-disputed) telephone – but it never caught on. In a world lit by sunbeams and electricity, ordinary light beams are too hard to send any distance without losing the message they're trying to convey. Even so, Bell's genius anticipated not just the mobile phone (which does a similar job with radio waves), but fibre-optic cables. These use laser beams to send digitally coded sounds and computer data down hair-thin glass or plastic 'pipes', ensuring the signals can't escape or degrade along the way.

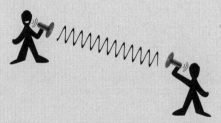

▲ **How Bell's photophone carries sound.** *The photophone carries sound on a beam of light. When you speak into the horn, the sound waves coming from your mouth make a black-and-white grating rattle back and forth. Inside the transmitter unit light shining from a bulb whistles through this shaking grating and flickers with the ups and downs of your voice. The light zooms through the air until it arrives at the listener. Inside the receiver unit an electronic detector converts the light rays back into electrical signals, powering a loudspeaker that recreates the sound of the speaker's voice.*

Radio days

Most modern forms of communication – satellite navigation, mobile phones, TV and Wi-Fi Internet – draw inspiration from the work of Heinrich Hertz, the real father of radio. While Italian showman-entrepreneur Guglielmo Marconi generally gets the credit for *inventing* radio, he was merely its populariser. But for an accident of history, he might not even have been that. When he left Italy in 1896 to demonstrate his radio prototypes to, among others, the head of the British Post Office, he was stopped from entering the country by customs officers who suspected his mysterious equipment was an elaborate bomb.

The theory of radio came from playful Scottish physicist James Clerk Maxwell. Knitting electricity and magnetism together in four simple mathematical equations, he produced the first combined theory of electromagnetism in 1873. The basic idea is that electricity and magnetism aren't totally separate things, the way they're taught in school, but more like flipsides of the same coin (you never get one without the other). Hertz expanded this theory and, 14 years later, sent the first proper radio-wave transmission. As we saw in the previous chapter, radio waves are simply a different kind of light – longer in **wavelength** (the distance between one wave peak and the next), shorter in **frequency** (the number of times the wave peaks each second), but otherwise no different in kind. Hertz was the man who proved this for sure in a crackling, fizzing German laboratory, the very same year that Albert Michelson and Edward Morley, two American physicists, finally busted the idea of the aether. Radio was on a roll.

As Marconi ably demonstrated, radio waves were perfect for beaming sounds over long distances. When others discovered how to send images alongside them, television was born. Farm boy Philo T. Farnsworth, America's answer to John Logie Baird, got the idea of making a TV picture (by scanning parallel lines of light) while ploughing one of his father's fields into perfectly neat furrows. Never properly recognised as the true father of television, Farnsworth came to despise the way his lofty, educational invention descended into something more tacky and tainted – the 'end of the pier' in a corner of the sitting room – and squandered his life as a penniless alcoholic[155].

Meanwhile the reach of radio didn't stop with TV. Radio waves could be bounced off things to see how far away they were or how quickly they were moving, which gave us radar. Now we can shoot radio waves up to satellites and back down to Earth to carry telephone, TV and Internet signals from one side of the planet to the other – all in a matter of seconds. With the exception of fibre optics (zapping information down wires using beams of laser light), virtually every modern form of long-distance communication involves blasting radio waves from one place to another – but how exactly does that work? How do Albanian pop stars find their way inside your TV set, in a fraction of a second, using nothing but vibrations of empty space?

How your radio really works

Suppose you want to talk to a friend in Africa and all you have at your disposal is a single electron – one of those 'fly' particles buzzing around the empty cathedral of an atom. Could you do it? Theoretically, yes. When electricity moves back and forth it creates magnetism. You might have done that little trick where you put a compass needle near an electricity cable and watch it move. That happens because the electric current, surging back and forth in the cable, generates a magnetic field all around it, which is what tricks the needle into turning. By the same token, when magnetism fluctuates, it produces electricity. If you pedal a bike wheel with a dynamo attached to it, what you're really doing is making a coil of wire spin inside a magnet. The magnetic field in the wire constantly fluctuates, producing electricity that lights your bike lamp. With the exception of solar energy, virtually all the electricity we make comes from electromagnetic generators in this way. So what about radio waves?

Make your own radio station

Suppose you take your electron and shake it up and down very quickly like a jar of reluctant ketchup. It has an electric charge, so moving it back and forth generates a magnetic field. But the magnetism fluctuates and a changing magnetic field generates electricity. So moving a charge up and down produces simultaneous, interconnected electric and magnetic fields that egg one another on. These waves of undulating electricity and magnetism ripple outwards from the moving electron and race away at the speed of light. This is what we really mean when we talk about radio waves.

In practical terms, the best way to go about this is to make your electron vibrate up and down along a straight metal rod – better known as a radio **antenna**, or aerial. Just as you can turn vibrating charges into an outgoing radio wave with an antenna (a transmitter, in other words), so you can turn an incoming radio wave back into electric charges, signals and sounds you can hear with the help of a second antenna (a receiver). As a good rule of thumb, an antenna has to be about half the wavelength of the waves it's transmitting or receiving. In the case of a mobile phone, operating on a frequency band of about 2 GHz (2,000,000,000 hertz), the microwaves that carry calls are about 15 cm (6 in) long, and the antenna you need is about the length of your little finger (either pulling out from the handset telescopically or, in newer phones, cunningly concealed inside the case). FM transistor radios operate on lower frequencies than mobile phones (and therefore have longer wavelengths). Like older mobile phones, they use pull-out telescopic antennas typically about 1–1.5 m (3–4 ft or so) long. Do the sums and you'll find that's roughly half the wavelength of a typical FM broadcast[156].

Distance no object

Like light waves, radio waves fire in straight lines. If that was all they could do, radio transmitters would be of little more use than lighthouses. Their signals would simply shoot off into space, making them no good at all for sending messages more than about 15–30 km (10–20 miles)[157]. What makes them so handy for communication is that they can bend around our curved Earth very easily. This happens for two very interesting reasons. First, if a tall radio mast is grounded (connected to Earth), the planet itself behaves like the bottom half of the antenna. Imagine a mast standing on a lake with perfectly still water beneath it. What you'd see, looking from the side, would be a mast twice as tall: the antenna itself standing on its own reflection in the lake. The same thing happens to a radio mast that's earthed: the planet conducts electricity so that it acts as a natural, mirror-image extension to the mast. As a wave spreads out from an antenna, it naturally curves around the contours of the planet in what's known as a **ground wave**.

▼ **How a transistor radio works.** *A transistor radio catches incoming radio waves whistling through the air. When the waves (rippling patterns of electromagnetism) hit the antenna (aerial), they make electric charges vibrate up and down it. This creates an electric current that the radio's circuitry turns back into music or speech that you can hear. The incoming signals are normally very weak; the transistor is the part of a radio that amplifies signals so that they are strong enough to drive a loudspeaker.*

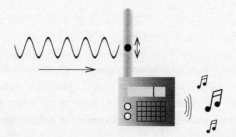

The other reason why radio waves travel so far is even more clever. As most people are well aware, if you use an AM (medium wave) radio, you can hear all kinds of crackling, buzzing foreign radio stations at night that you can't detect during the day. That's all down to a part of Earth's atmosphere called the **ionosphere**, roughly 60–500 km (40–300 miles) above your head (at least six times higher than jet planes fly). It has that name because it contains ions (atoms split apart into positively charged bits by losing some of their negatively charged electrons), which means that it conducts electricity. The ionosphere is dramatically affected by radiation from the Sun, so the way it behaves changes radically between daytime and night. During the daytime the lowest layer of the ionosphere soaks up radio waves and prevents them from travelling very far. At night that effect diminishes, enabling higher layers of the ionosphere to reflect radio waves like a mirror, intercepting signals that would normally shoot out into

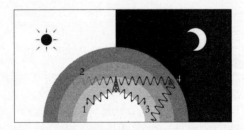

▲ **How AM radio waves travel differently in the day and at night.** *In the daytime, ground waves leave an antenna and follow the curvature of the Earth (1). Waves shooting off towards space (2) are absorbed by lower layers of the ionosphere and get no further. At nighttime the ground waves travel in the same way (3). However, the lower layers of the ionosphere no longer soak up and block radio waves. Instead, the waves penetrate through to the upper layers of the ionosphere, which bounce them back down to the ground, making it possible for signals to travel much further (4).*

space and shining them back down to the planet's surface. Some radio waves repeatedly bounce between the ground and the ionosphere, effectively carrying them from one side of the planet to the other.

Mirrors in space

The cunning little trick of bouncing radio waves off the sky delighted and divided the greatest minds of the day. When Marconi fired radio waves 3,200 km (2,100 miles) from Poldhu, Cornwall, in Britain, to Newfoundland in 1901, Alexander Graham Bell refused to accept that he'd done so. He said, 'I doubt Marconi did that. It's an impossibility,' although Thomas Edison was more open-minded, stating, 'I would like to meet that young man who had the monumental audacity to attempt and succeed in jumping an electrical wave across the Atlantic[158].'

The explanation – the theory of how the ionosphere reflects radio waves – was an inspired insight from the mind of Oliver Heaviside, a first-rate English physicist whose name is largely unknown today outside academia[159]. As a young man he revolutionised global telecommunications; in later years, as a recluse, he plunged into a downward spiral of eccentricity. He swapped his clothes for a kimono, threw out his furniture and replaced it with granite rocks, papered the walls of his cottage with unpaid gas bills, lived on milk and cookies, and took to firing off polite letters of complaint to errant neighbours signed 'His Wormship, Professor Oliver Heaviside (W.O.R.M.)'[160]. It's fascinating to think that the progress of science and technology hinges, so often, on brilliant people who ride the divide between genius and madness. Was it charming eccentricity adding colourful zest to dull academic life, or tragic mental illness robbing the world of further amazing insights? We'll never know. His good friend George Searle, another notable physicist, described Heaviside as 'a first-rate oddity', but 'never a mental invalid'[161].

It was probably just as well that Heaviside had already shared this cunning secret of the sky, because when he predicted how the ionosphere could help to transmit radio waves, in 1902, he paved the way for the modern age of global radio and TV broadcasting. What could be better than having a mirror high in the sky, helpfully bouncing radio signals around the world? The trouble with the ionosphere is that it's a natural phenomenon and its ability to help us relay signals varies dramatically at different times of day and even with the weather. Wouldn't it be so much better if there were an actual mirror in the sky that we could use to communicate more reliably? That's the basic idea behind **communications satellites**.

Although other people had the same idea earlier, science-fiction author Arthur C. Clarke is generally credited with inventing the idea of a stationary spacecraft that can help us to send messages from one side of our planet to the other, exactly like a space mirror. In his original proposals, written in 1945, he suggested pinning three satellites in precise orbits, effectively static above different parts of Earth, as the planet whirled around beneath them[162]. That meant they had to be 36,000 km (22,000 miles) above the ground, moving in paths that are now known as **geostationary orbits**. The idea remained highly speculative and theoretical until 1957, when the Soviet Union successfully launched *Sputnik 1*, the first artificial satellite. It wasn't a communications satellite and it wasn't in geostationary orbit, but it was a big step in the right direction. Three years later the United States moved another step forwards with *Echo*, the world's first communications satellite, which was literally a mirror in space.

Unlike modern communications satellites, which are truck-sized metal cans packed with electronics, powered by unfolding solar panels, and costing hundreds of millions to develop, *Echo* was a giant, inflatable Mylar plastic balloon approximately 30 m (100 ft) in diameter. It was a simple proof of the concept that allowed scientists to boot a radio wave up into space and have it bounce back down to the ground – just like a ball kicked against a wall. After that successful experiment, the first communications satellite, *Telstar*, was launched a mere two years later. By 1965 the world had INTELSAT 1 (*Early Bird*), the first true geostationary communications satellite, which could successfully relay 240 separate telephone calls or a single black-and-white TV picture. Vastly more sophisticated, modern communications satellites can carry hundreds of TV channels at the same time.

Why use radio waves?

If all forms of electromagnetic radiation are really just the same stuff, why do we use radio waves to communicate and not, for example, X-rays or gamma rays? After all, they have super-tiny wavelengths so, instead of needing giant antennas to communicate, wouldn't a radio receiver that used gamma rays need an equally tiny (and much more convenient) transmitter and receiver?

The answer is two-fold. First, the wavelength of electro-magnetic radiation is opposite in size (or, as mathematicians like to say 'inversely related') to its frequency. The longer the waves, the smaller their frequency (and energy). In the case of gamma rays and X-rays, although the waves are tiny (the wavelength is atomically minuscule), the frequency is extremely high. We measure radio waves with frequencies of megahertz, but X-rays and gamma rays have frequencies (and therefore photon energies) about a trillion times higher. In quantity, they are harmful to human health: the last thing we'd want is to bombard our homes with what are, in effect, atomic death rays.

Some waves with longer wavelengths are also harmful. Microwaves spring to mind. Microwave ovens have metal-lined 'cooking cavities' designed to stop harmful radiation from escaping, yet the 13 cm (5 in) waves used inside them are much the same as the microwaves used to fire signals between mobile phones. The main difference is that microwave ovens use several thousand times more power. Although there's an ongoing (and very important) debate about whether mobile phones truly are safe to use, this crucial difference alone suggests that microwaves will happily sizzle your chicken korma while safely leaving your cerebral cortex unscathed[163]. Put an active mobile phone on top of a TV dinner inside your microwave and, if it cooked at all, it would take several thousand times as long – perhaps eight days instead of four minutes[164].

The other reason why we use radio waves is that they travel further. Like the footsteps of a giant bounding through the sky, they're big enough to dodge buildings, houses, trees and cars, usually with little loss of signal. You can pick up radio stations inside your concrete-block home or your metal-lined car, no problem at all. With the help of long-wavelength radio waves, the BBC World Service can wiggle and wave 'God Save the Queen' right around the planet – to all the scattered fragments of the faded British Empire. But if the same transmitters were somehow reconfigured so that they beamed out X-rays instead, they'd have a job getting signals even to the next street. Ultra short-wavelength X-rays are much more readily absorbed by everyday objects and can't get through heavy metals like lead.

God speed, light speed

It's been a long journey from smoke signals to space satellites, but the whole process has come full circle: early forms of communication nipped from person to person at the speed of light – and they still do. The difference is that the people involved no longer need to be within actual *sight* of one another, and the 'light' that powers their conversation can (theoretically) be any kind of electromagnetism. There's nothing better than light for sending information, because there's nothing *faster*. But why is that so? What's so special about light?

We think of light purely in terms of illumination. The Sun lights our lives for half the day; electric lamps take care of the night. Light powers our vision but, even if we're unlucky enough to go blind, we know we could gradually learn to live without it. In short, light is nice to have but (leaving aside the power it provides to make our food) almost incidental. The world doesn't stop when we close our eyes. If, however, you think of light as endless waves of electricity and magnetism, bolting through space, or if you consider Einstein's own equation $E=mc^2$, in which c (the speed of light) is the missing link between energy (E) and matter (m), it's obvious that 'light' (electromagnetic radiation) is something much more vital. Talking on the phone or listening to the radio makes an intimate connection with the hidden, atomic universe. The sunbeams streaming through your windows, the radio waves shaking up and down the antenna, the satellite links pinging your voice to friends in far countries – all these tap into the very essence of matter.

Why is light so fast? It's a meaningless question. Light seems quick to human-sized creatures parked on a 'pebble' in the Milky Way. It wouldn't seem so fast if you were on the opposite side of the Universe waiting 100 billion years for your email to arrive. If you think hard enough, the question soon turns from physics to philosophy, from 'Why is light so quick?' to 'Why does light exist?' or 'What is light?' – and none of us will ever have an answer to a mystery like that.

CHAPTER TWELVE

Living by Numbers

In this chapter, we discover

> *How you can hold an entire bookcase with one fingernail.*
> *Why it could take you an hour to stroll across a compact disc.*
> *Why it's always better to buy a CD than download an MP3.*
> *Why digital pirates will always win in the end.*

Howard Hughes – the patron saint of recluses – would have relished modern life. Just consider: you can watch *North by Northwest*, download the latest *Harry Potter*, video-chat with your sister in Hong Kong, order kippers and cornflakes for breakfast, and reply to your work emails, all without leaving home, without even stirring from bed, if that's too much. What makes all this possible isn't the Internet, or even the computer, but something much more fundamental. The modern world revolves around a single, simple but terrifically powerful idea: every form of information, from knowledge and culture to news and emotion, can be turned into long strings of numbers and zapped from one side of the planet to the other at the speed of light.

If you're old enough to recall the 1970s and '80s, before digital technology first started to change our lives, you'll remember a very different world. Maybe you had a hefty library of skinny vinyl (*Astral Weeks*, *Tubular Bells* and *Trout Mask Replica*), or a bookcase packed with the cracked spines of titles such as *The Tin Drum* and *Catch-22*. Your boozy student photos were most likely tacked down in albums, each snap carefully arranged and glued to the page with four triangular cover slips. If you scribbled secret thoughts in a diary or notebook, you probably stuffed it in your sock drawer where no one could find it. Perhaps you had a filing cabinet or a box file where you jammed those stuffy letters from the bank and other vital papers you knew you had to keep.

How things have changed. These days your music collection is more than likely crammed onto an iPod slimmer than a pack of cards. You probably still have bookcases, but may also have the electronic equivalent on a Kindle or Nook. Family photos? Those are quite likely to be on your computer too, perhaps uploaded to a photo-sharing website such as Flickr or Instagram, or tweeted to the world the minute you snap them with your phone (the easy ability to share photos often now motivates us to take them in the first place). Letters to friends? Stripped of handwriting swirling like honeysuckle, for many of us those have evolved into anonymous, clinical emails filed away in a program like Hotmail or Gmail – floating up in 'the cloud', who knows where in the world. Think Facebook and Twitter, wave goodbye to privacy and open your sock drawer wide to the digital world.

The digital difference

How tall are you? Suppose you mark off your height with a pencil, childlike and upright, straight against the wall. The vertical length of plaster between the floor and the high line you make is a representation or 'analogy' of your height. It's a stand-in, if you like, for your *standing* – what scientists would call an **analogue** measurement. It's interesting (especially for children, if you record their height every few months while they're still growing), but not all that useful. If someone wants to know your height, what do you gain by showing them a mark on the wall? They can tell your height just by looking at you.

It's much more interesting and far more useful to turn an analogue measurement like that into its **digital** equivalent.

Using a ruler you can convert the vertical length of wall into a number such as 183 cm (6 ft). Digital measurements are much easier to compare. Am I taller than you? We could stand back to back and check, but we'd need to be in the same room to do it. It's quicker and easier just to swap our numbers. If you want to buy a coat online, for example, knowing the (digital) measurements of your body and comparing them with the size of the clothes in stock is much easier than staring at a scaled-down photo and trying to guess the fit. What goes for the outside of your body applies just as much inside: quite a bit of modern medicine is based on making digital measurements of things like our blood pressure or heart rate, and comparing them with values we know to be sensible and safe.

Suppose you go through this exercise comprehensively, measuring every aspect of your body you can think of, and turn yourself into the kind of technical data you'll have seen in a car catalogue or hi-fi brochure. How much do you weigh? What's your inside leg measurement? What's your IQ? With no imagination whatsoever, it's easy to reduce yourself to a simple spec that quickly gives the *measure* of you. What you'll effectively have done is mimic the process used by mobile phones, digital cameras, CD players, digital radios and computers: you'll have converted yourself from analogue form to digital form. If online dating's your thing, you can probably see how handy it is for a computerised matchmaker to compare people across a variety of dimensions using a simple string of digits.

It's not just in love that digital technology rules our lives – you've only to travel by train to witness many people around you either chattering into a (digital) mobile phone or pawing away at its screen – but it has drawbacks as well as benefits. What are you actually like? Can you sum yourself up in nothing but numbers? And, if the answer is 'no', why do we suppose we can compress a Picasso painting into a digital photo or squash a Beethoven piano sonata into an

MP3 file? What do we gain when we live by numbers – and
how much we do lose?

Machine law

We're analogue animals: we hear sounds, see pictures, feel
emotions – none of them easy to turn into words, never
mind numbers. Computers are digital devices: whatever
they're doing, they're processing, ultimately, in digital form.
Not a human-based, numeric representation using the 10
characters 0–9, but a boiled-down, **binary** equivalent: a
computer's ability to reduce any ordinary number (or any
other information) to zeros and ones. A computer converts
a relatively palatable number like 12345 into 11000000111001
so it's easier to store and process with electronic switches
that can flip between only two states, on (representing the
number one) or off (representing zero). These days the most
powerful chips pack over 2 billion of these switches, which
are called **transistors**, into a space the size of your fingernail.
If it takes eight transistors to store one letter or number, that
space is, theoretically, good for storing 250 million characters
and 400 fairly sizeable books – enough to fill a modest
home bookcase.

The difference between analogue humans and digital
devices is even more subtle than it seems, because people
can't help but pin meanings to the meaningless. The number
above, 12345, is a meaningless example that nevertheless
means something: it's an easy run through the first five
digits – a digital equivalent of 'ABC'. Or how about a price
tag in a shop? When you look at a figure like £49.50 or
$145.67, your brain instantly converts that by association
into much more than just a number. You relate it to other
things that cost more or less, what else you could buy for
the same amount or how long it would take you to earn. We
remember randomly assigned ATM machine (cashpoint)
PINs by converting them into birth dates, or pick lottery
numbers using numerically coded wedding anniversaries.

Our brains insist on finding meanings wherever they look – and put them there if they can't.

Computers, by contrast, find no meaning in numbers they crunch like cornflakes. They can't distinguish between a string of digits that signifies a brush stroke by David Hockney and an identical numeric sequence in a sound file that represents 'Gimme Shelter' by the Rolling Stones. But that's only the half of it – and the better half at that. The real problem is that computers are digital philistines; to a computer, the deepest human meanings are void of *all* meaning. In a computer's binary mind, there's no distinction whatsoever between the tweets of a celebrity and the Quran.

The analogue age

Not that the vacuity of computers has anything to do with their obsession with numbers. Before digital technology arrived in the mid-1940s, the best computers were entirely analogue. They chewed through artillery calculations (How far would a bullet fly? What if the wind were blowing twice as hard?) with the gnashing teeth of oiled cogs, measuring their results by how much a gear had ground around or a marker had shuffled down from its original position.

Some of the finest analogue machines were developed between the two world wars by US government scientist Dr Vannevar Bush, later a key figure behind the atomic bomb and the grandfather of hypertext, the click-on-a-link technology at the heart of the World Wide Web. Bush's best machine, the Rockefeller Differential Analyzer, was a 100-tonne (110-ton) beast that filled a room, built from 320 km (200 miles) of wire, 150 electric motors and 2,000 vacuum tubes (the forerunners of transistors). It looked like a giant pinball machine, but the only games it played – for the US military – were ones of life and death.

Fittingly, this analogue avenger was rendered obsolete by an entirely *digital* machine: the infamous ENIAC (Electronic Numeric Integrator and Calculator), completed in 1946.

ENIAC was still a beast, though somewhat more compact than those built by Bush. Weighing some 30 tonnes (a third as much), it filled a room 10 × 15 m (30 × 50 ft), had around 100,000 electronic components and five times as many soldered joints, and gobbled as much electricity as 60 electric toasters burning bright all day and night[165]. But it could really sing. Where it took a human 'computer' (a mathematician working with a slide rule) about 40 hours to crunch through a single artillery calculation, Bush's Analyzer could do the job in 30 minutes. ENIAC cut that spectacularly: it could do up to 5,000 simple additions or subtractions per second. Its first notable success was chewing through a nuclear physics problem estimated to need 100 man years of human calculation. ENIAC cracked it in a mere two weeks (two hours for the actual calculation, but a further 14 days for programming, analysing the results and perhaps just a minute or two of self-congratulation)[166].

HOW DOES DIGITAL TECHNOLOGY WORK?

From mobile phones and iPods to calculators and computers, the trick at the heart of any digital device is converting analogue to digital and back again. That's easy enough when we're talking about a person's height or weight, and there's only one simple quantity to represent. But what about 'measuring' the *Mona Lisa*? How would you convert that into a digital image you could store on a computer?

First off, you wouldn't turn it into one number, but millions. The process is called **sampling**, and it means dividing a lump of information into small chunks, measuring each chunk, turning the measurement into a number and stringing all the numbers together. So we could divide the *Mona Lisa* into maybe 1,000 rows and 1,000 columns, or a

▷

million individual squares. We could measure the average colour and brightness of each square (giving both a number), and run those measurements in sequence from left to right and top to bottom. That would turn one picture into a pattern of 2 million numbers (or one number consisting of 2 million individual pieces, if you prefer), which would be relatively easy to store in a computer or rattle down a phone line. An analogue picture becomes a map-like grid-pattern of on-off binary digits (or bits) – a bitmap, as it's technically called.

Let's see how sampling works in digital cameras and MP3 music players.

Digital cameras

An old-style camera has a lens and shutter that opens briefly to let light into a sealed compartment, where it 'exposes' a piece of plastic film coated with silver-based chemicals. Light turns this silvery stuff into tiny bits of silver, which clump together, so bright parts of the scene produce dark areas on the film and vice versa. In other words, a photo starts off with its black-and-white areas reversed in what we call a negative. When the negative is printed, the image is inverted. Dark bits become light, light bits turn dark – and the final 'positive' print recreates the original scene.

A digital camera works in a completely different way, using a light-sensitive chip called a **CCD** (charge-coupled device). Unlike an exposed piece of film, which becomes a continuous *analogue* representation of the object, a CCD is divided into millions of light-sensing 'compartments', or pixels, each of which measures the light falling onto it and outputs a number. So a CCD automatically transforms an analogue image into a digital photograph.

▲ **How image sampling works.** *Here I've sampled the Mona Lisa at five different resolutions, and it's quite obvious that the more pixels I use, the better the image becomes. The extreme left image is made from just 10 × 15 = 150 pixels, but if I asked you to guess what the picture was, and you squinted and relied on your memory (and the likelihood of my asking about an iconic image), you could probably guess correctly. The second left image has 20 × 30 = 600 pixels (four times more) and is easily recognisable if you squint. The other images contain progressively more pixels, but don't give us proportionally more information. For example, the image in the centre is 40 × 60 = 2400 pixels, while the image on the extreme right is 600 × 900 = 540,000 pixels (0.5 megapixels) – about 225 times more. Even so, it doesn't give us 225 times more useful information than the picture in the centre. That's why sampling works so well: our brains can fill in the missing bits. The mathematical magic trick of squeezing an image into fewer pixels is called compression. Because it throws away extra information (which can never be recovered), it's known as lossy compression (the image files taken by digital cameras, JPGs, are examples). When you select different levels of quality for saving photos, you're simply choosing different levels of compression.*

MP3 players

A digital camera makes photos by using a CCD to sample space (the area in and around the object you're snapping). By contrast, the recording equipment that makes an MP3 music file samples sounds by measuring them repeatedly over a period of *time*. Imagine you want to make a

▷

downloadable MP3 from your old vinyl copy of 'Gimme Shelter'. You could play the record on your speakers with a microphone propped in front of them. Plug the microphone into your computer and it will pick up the *analogue* sound waves the speakers play. With appropriate encoding software and electronic circuitry, you can measure those sounds roughly 44,000 times per second and convert them into numbers. An MP3 or M4A file, such as the music tracks you buy online from Amazon or Apple's iTunes store, is just a long string of these numbers.

Never mind the quality?

Analogue or digital, a copy of something can never be as good as the original – or can it? When digital photos were still quite new, many people thought old-style (analogue) film photos were much higher in quality – and they were absolutely right. Now it's impossible to tell the difference, and even professional photographers have switched allegiance. That's because new-style digital cameras use far greater sampling than their older counterparts: their CCDs have millions more light sensors – millions more pixels, in other words. A digital camera boasting 10 megapixels has a CCD sensor that converts an image into about 10 million measurable dots. Compared to an older camera with only a 2 megapixel CCD, making a picture exactly the same size, that's a five-fold increase in sharpness and detail. Interestingly, a quick search online suggests that old-style photographic film gives a resolution somewhere between 10 and 50 megapixels (though different kinds of film and the developing and printing process might reduce the quality somewhat), so even an average digital camera still, theoretically, fails to rival the extreme quality of old, analogue SLRs, even if our eyes can no longer really tell the difference[167].

The same is true of MP3 music files. The more often you sample and measure the frequency and volume of the sound, the closer your digital recording will be to the original piece of music and the higher the quality. The catch is that the more often you sample, the more measurements you make and the bigger the digital file you end up with. That's why higher quality MP3 music files sampled at higher rates are larger, take longer to download and fill up your iPod more quickly.

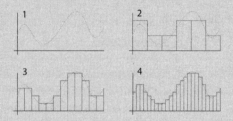

▲ **How music sampling works.** *Suppose we have an original, analogue sound lasting about six seconds (dotted line in graph, 1). If we want to turn it into a digital file, we need to sample the sound: repeatedly measure the height of the graph and turn each measurement into a number. If we sample every second, we need to make six measurements (2), but our digital approximation to the original sound is very crude (block graph). What if we measure twice as often? We get a more accurate approximation (3), but we have twice as much data to store (12 measurements represented by 12 blocks). If we sample twice as often again (4), we get an even more faithful representation of the original sound, though some of the detail is still missing. This time we need 24 measurements – and the digital music file we create will be four times bigger than the original, crude sample directly above it. In short, there is a trade-off between the quality of a sample and the size of the file you need to store it.*

What's so nice about numbers?

Why the sudden craze for digital technology? Do we digitise so much information simply because we can? And if so, why?

There are all kinds of advantages to going digital. If you're making a mobile phone call your words travel as numbers, giving sharper sound, because digits are easier to send and receive without getting jumbled on the way. Digital mobile calls are also encrypted, making it impossible for eavesdroppers to listen in and giggle at the things you're saying. With old-style analogue mobile phones, you could intercept calls whistling through the air using a simple piece of equipment called a scanner. Not such a worry for most of us, but critically important to spies, adulterous actors and shifty politicians[168]. Another obvious benefit of digital information is how little space it occupies. You can fit somewhere between 1,000 and 2,000 ebooks onto a typical ebook reader (equivalent to maybe 40 shelves of paperbacks or about five well-packed bookcases). If we assume, for the sake of easy maths, that the average Kindle can store 1,500 books, the whopping 36 million books that pack the vast vaults of the US Library of Congress would comfortably squeeze onto 20,000 slim Kindles, which you could happily stack in one room (200 stacks, 100 Kindles high)[169]. Of course, a Kindle is mostly screen, battery and plastic case – the all-important memory chip takes up a fraction of the space inside. If you really wanted to, you could squash all that information onto a single 40 terabyte (40 trillion byte or 40,000 gigabyte) memory drive the size of a large briefcase.

Music is equally impressive, or perhaps even more so: thanks to 'lossy' MP3 compression, you'll certainly fit 500 CDs onto a typical iPod you can pop in your pocket (laid end to end, they'd stretch the length of a family car). You can do the same trick with photos: around 500–1,000 digital photos will comfortably squeeze onto an SD memory card the size of a postage stamp. (The photographic films in old-style film cameras were limited to just 24 or 32 snaps; digital cameras have completely changed photography by making it possible for people to capture five, 10 or 20 versions of a picture, and keep on clicking away until they get the shot they really want.) The less space digital information occupies, the faster it is to zap across the Internet. You can email a digital photo to a friend in the blink of an eye and download an entire ebook or music album in much less than a minute.

Most telephone calls rattle, for at least part of their journey, down fibre-optic cables 5–10 times thinner than a human hair (each fibre carries about 20,000 calls at once, so the entire cable can carry millions of calls at a time). How is this possible? Because digital information can be squeezed and compressed in ways that are impossible with analogue. Where an analogue call has to transmit, faithfully, all the pauses and repetitions in the basic sound waves making up your voice, digital calls can smartly encode these things using compression, so they can be transmitted in a fraction of the space and time. You might have noticed how streaming TV programmes (watched over the Internet) jump and jerk a lot when the actors walk too quickly or the credits roll at the end. That's because they're downloading a highly compressed video – and the compression has been achieved by removing some of the information that makes the rapid action in a normal TV programme so smooth and seamless.

Another, less obvious advantage is that digitised information is easy to edit in all kinds of ways. Many of us have used computer graphics programs to touch up digital photos – or added words to images to turn them into greetings cards or Internet memes. While sampling can certainly compromise the quality of digital information, it's not always a drawback. Digital radio and TV don't generally suffer the snap, crackle and pop of interference that used to plague old-style analogue broadcasts.

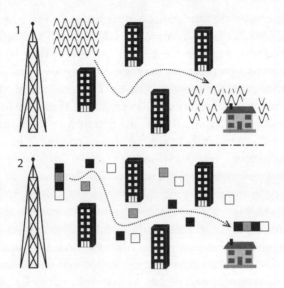

▲ **What's so good about digital radio and TV?** *With old-style broadcasting (1) the signal is boosted but transmitted from the mast much as it is – in waves that bear a strong resemblance to the original picture or sound. Some gets lost on the way, so the signal that arrives in your home is degraded by crackles and whistles. With digital TV (2) the signal is turned into a numeric code and split into chunks (black, white and grey squares), each of which is sent many times. Some chunks get lost, but enough arrive at your home to recreate the original signal pretty well. Digital TV isn't perfect, however. If enough chunks go missing, you'll lose the entire signal; with analogue TV, you might have to suffer a slow and steady degradation of the signal into crackle and snow, but you'll almost always pick up something.*

Digital dilemmas
That's not to say that digital technology doesn't have its downsides – indeed, there are plenty of them.

Herding copy cats
Copyright is an obvious issue. As the music industry has already found to its cost, once music is digitised and turned

into MP3 form, there's almost no way to prevent people from sharing it online for free. Music stores such as iTunes attempted to get around the problem with a two-pronged attack. First, by making digital music relatively cheap and freeing affordable individual tracks from more expensive albums packed with filler you didn't really want anyway, they undermined one of the main reasons for music piracy: if music is cheap enough, the theory goes, people won't mind paying for it. Second, and largely for the reassurance of the media multinationals, iTunes originally locked its music files with a special code to prevent them from being shared by too many people – a basic approach known as Digital Rights Management (DRM)[170]. Ebooks and downloadable films are protected in the same way.

In theory DRM files are encrypted so that only legitimate buyers can play them. In practice, although doing so is unlawful[171], it's a cinch to remove the encryption with hacked software that can be downloaded and installed in about five minutes. Conventional (newspaper pundit) wisdom sees the problem of piracy and online file sharing as an ongoing tug of war between staid, overpaid geeks working for the publishing multinationals, devising ever more sophisticated DRM systems, and spotty underpaid geeks reverse engineering their work and rendering DRM little more than cosmetic protection. Companies such as Google sit in the middle. They happily comply with their legal duty to remove pirate material, providing the copyright owners bring the issue to their attention, but they could never be described as proactive: most pirate material stays on a site such as YouTube indefinitely until and unless the copyright owner bothers to complain.

In practice it's impossible to protect any kind of digitally encoded information from piracy. Why? Because you can always turn it back to analogue information, encode it digitally without DRM and distribute it yourself in unencrypted form. Legality and morality aside, there's nothing to prevent you from retyping an ebook into your computer and producing a file you can share, illegally, with your friends. In this case your brain and fingers are doing the digital to analogue conversion. It's even easier with music. You can take the most sophisticated, digitally encrypted music file, play it through your computer's sound card, catch the analogue signal, route it back through an MP3 encoder and build yourself a beautifully unencrypted MP3 file that sounds just as good as the original MP3. It'll take you as long as it takes to listen to the track plus a minute or so to do the encoding – maybe five minutes in total[172].

What about the quality?
Often, pirate copies don't sound perfect, because MP3s themselves fall short of perfection. As we saw in the diagram above, a sampled digital music file is usually an *approximation* to the analogue track from which it was recorded – and there's usually something missing[173].

You can get a sense of just how approximate an MP3 is by comparing it with an identical track on a CD. Compact discs store music in the form of microscopic bumps called pits, curling around in a spiral that systematically unwraps (in the opposite way to an LP record) from the centre of the disc to the edge. A typical CD contains about 3 billion of these pits, each about 600 nanometres wide (a couple of thousand times wider than an atom) and three times longer, stamped in a line that, fully unwrapped, would stretch to about 6 km (3.5 miles); it would take you about an hour to walk from one end to the other. If there are 10 tracks on a CD, that means there are 300 million pits per track. For simplicity, let's assume the CD is encoded so each pit is equivalent to one bit. That means we have 300 million binary digits to encode perhaps four minutes' worth of high-quality sound. Does that sound reasonable? Who knows? But consider this. Each track on a CD takes up about 60 megabytes. The same track converted into MP3 format takes up only about 6 megabytes, so there's a 10:1 reduction in the amount of digital information when you 'rip' a CD and produce an MP3. An MP3 album is a highly compressed version of the digital tracks on a CD, which are themselves only an approximation of the original analogue sound. That's why you're always better off buying the CD of an album and ripping your own MP3s than buying the MP3 version for a similar price: you'll always have a higher quality recording at your disposal, should you need it[174].

A fragile permanence

There are 48 priceless copies of the original Gutenberg Bible still in existence, each one lovingly handprinted in the 1450s, almost six centuries ago[175]. But no one has a record of the very first mobile phone call, made in 1973, the first email message, sent by Ray Tomlinson in 1971, or the first SMS text message[176]. A vast amount of digital data produced in the 1980s and stored on old-fashioned floppy discs or magnetic tapes has long since disappeared into landfills – and even the worms have lost interest.

My favourite example of this is the BBC Domesday project from the 1980s. In an attempt to commemorate the 900th anniversary of the original Domesday Book from 1086 (a historic document recording land use and ownership in England and Wales), the BBC invited 14,000 British schools to compile, for posterity, a meticulously detailed record of their local communities. A hugely impressive feat, it netted a gigantic database with more than 148,000 pages of text and 23,000 photos. Unfortunately, the BBC decided to publish the results on proprietary laser discs (the over-sized, high-capacity video CDs predating DVDs), which rapidly became obsolete. Today, the only place where you can read the original 1086 Domesday Book is the UK National Archives in London. Until recently, thanks to the limitations of failed digital technology, that was pretty much the only place where you could read the BBC's 1986 equivalent. Fortunately, some of the project has now been saved and relocated to the Web[177].

If that makes you smile, consider for a moment your own experience of digital communication. You might keep cherished personal letters or birthday cards from friends or relatives, but how many emails do you save routinely – and would you ever want to? According to a survey by the Radicati Group, on average business people send and receive 110 emails every single day, which translates into a million or more over a working lifetime[178].

Information is certainly more throwaway and ephemeral than it ever was in the analogue age. We might still think before we speak, but do we engage our brains before we email, text or tweet? Will anyone really care if the vast binary chit-chat of the 21st century disappears into nothingness almost as quickly as the people who created it? If talk is cheap, digital talk is cheaper, more disposable and far more meaningless. Modern life might be digitally dictated, but is it necessarily better that way? Perhaps living by numbers is as true a reflection of real life as painting by numbers is a meaningful reflection of art.

CHAPTER THIRTEEN
Blowing Hot and Cold

In this chapter, we discover

> *How you can heat your home with nothing more than a candle.*
> *Why you can't cool a house down as easily as you can warm*
>   *it up.*
> *How you can make instant ice cream.*
> *How much heat energy a typical home holds in the middle of winter.*

If you live in a temperate part of the world, such as northern Europe or the East Coast of the United States, you probably spend half the year grumbling about the cold and the other half moaning about the heat. Such are the seasons, brought to us by a sorry stoop in Earth's axis that makes different latitudes heat up or cool down, in relatively gentle alternation, as our planet shuffles on its annual perambulation around the Sun. Most of us find the seasons a welcome variation: who'd want year-long snow drifts or heat waves? Yet the practicalities of keeping warm or cool are more problematic. We suffer the seasons in silent amusement because they're sufficiently spaced out. There's a less extreme hot-cold variation between night-time and day – and we suffer that much more easily, precisely because it is less extreme.

But what if these two things were somehow intertwined? What if midday were always midsummer and midnight always midwinter, and the swap between them happened every 12 hours? Then you can be sure we'd really be blowing hot and cold about the seasons: the rapid-fire switchbacks would soon send us crazy. Unlike our current predicament, lurching annually between hot and cold, we might finally wise up to the stupidity of heating our homes one minute and cooling them the next. We'd quickly develop better ways of storing solar energy or insulating against excessive temperatures so that the predictable fluctuations would bother us less – and cost less in heating, air conditioning and ventilation.

Or would we? Even inside our homes, we blow hot and cold. In your very own kitchen, you'll find a couple of super-cooled metal boxes (the fridge and freezer) within spitting distance of two superheated equivalents (the cooker and microwave oven). You'll find isolated pockets of hotness or cold (a kettle here, an ice-cream maker there, a washing machine that boils your sheets, and a tumble dryer that turns them to dry cloth and steam). In other rooms around your home, you'll find more of the same: electric fans and fires, power showers, curling tongs, irons, hairdryers and air conditioners.

Wherever you look, we're either heating things up or cooling them down. With copious electricity and gas, and plenty of money to pay for them, it's easy enough to turn hot into cold or vice versa. But why do we have to do it at all? Why don't warm winter homes or cool summer ones stay that way? Why this endless temperature tango? Why are we constantly blowing hot and cold – and what, if anything, can we do about it?

Unbreakable laws

Everyday laws are the bargain we strike for a life of tolerant peace and quiet: behave moderately, abide by social conventions and you'll be left alone to do as you please. But there's always a choice. You can opt to break social laws for personal gain at the risk of punishments that range all the way from a few sniffy neighbours to the sizzling shock of the electric chair.

The laws of physics are entirely different: they are absolute and uncompromising. We don't need scientific lawyers or courts, judges or juries, to weigh up our compliance and offer damning or reassuring verdicts, because these laws are black and white. There are no rewards for sticking to them and no punishments for breaking them because they're quite simply unbreakable. There's no physics equivalent of crime: breaking physical laws never gets further than a mind game (or what physicists like to call a 'thought experiment'). Even though liberals never fail to admire Bob Dylan's contrary observation ('To live outside the law you must be honest'), no such idealism is possible in physics: there's no life outside the law. Period.

What are these famous laws? Among the most important are a couple that explain why heat and cold (which is simply the lack of heat) behave the way they do. Known as the **Laws of Thermodynamics**, they offer an explanation for all our domestic blowing hot and cold – and much more besides. Thermodynamics simply means 'heat in motion', so its laws explain things like how cars waste energy, why power stations need such stupendous cooling towers, why cows have lovely damp noses and dogs dangle out their tongues – and even why Arctic musk oxen spend so much time standing still in the snow.

Hot on the inside

Heat is a kind of energy things have because the atoms or molecules inside them are jiggling around (with kinetic energy). Hotter things simply have more of this internal jiggling than colder ones. A gas (such as steam) is hotter and has more internal kinetic energy than the same amount of its equivalent liquid (water), which in turn is hotter and has more kinetic energy than the same liquid in solid form (ice). If you heat a gas the atoms or molecules inside it move faster, jiggle more and crash into one another more often. This way of picturing the heat inside things like a kind of internal, atomic game of dodgem cars is called the **kinetic theory** – and it can explain most of what we understand about heat and how it works.

Temperature is the thing most of us associate with heat energy, but it's a subtly different concept: a measurement of how hot or cold something is, rather than how much heat energy it contains. That seems confusing and unclear until you consider the difference between a hot cup of coffee and an iceberg big enough to sink the *Titanic*. Straight from the machine, coffee is probably at a temperature of about 90°C (194°F), while the iceberg might be -10°C (14°F) or even colder. The coffee, however, is a mere cup full of water. Even though it contains zillions of molecules, and their average energy is quite high (because the water is hot), they have only so much heat energy in total (we can estimate it by multiplying the average energy of each molecule by the total number of them). The iceberg might be vastly colder, but it's also vastly *bigger* – and size is what matters here. Even though each water molecule has less energy on average, there are so many more of them that the iceberg's total energy is far greater. The coffee is hotter, but a typical iceberg holds about 200 million times more heat energy[179].

Living within the law

If you stand your coffee on an iceberg two things happen. The coffee cools dramatically and (much less obviously) the iceberg warms up, albeit imperceptibly. Two things at different temperatures diplomatically *compromise*: they swap heat energy, bringing them to exactly the same temperature, until they reach an equilibrium. It's easy to see how this happens if you think about the kinetic theory. Hot water molecules crashing around in your coffee collide with the solid ceramic molecules in the cup, passing on some of their heat, cooling down in the process and heating up the cup itself. The cup, which is in contact with the cold ice, passes on its heat in the same way, heating up molecules in the ice and also cooling down slowly. So there's a kind of invisible conveyor belt of heat shipping energy from the coffee to the ice – and it keeps whirring away and transporting heat until both are at the same temperature.

As we saw in Chapter 2, energy is far from magic: it doesn't suddenly appear or disappear without warning. If something loses energy, something else gains it: energy exchange is always a zero-sum game. The same is true with our coffee cup and iceberg. The amount of heat energy that the coffee loses is the same as the amount the iceberg gains (assuming none is lost to the surrounding atmosphere). This, which we previously called the Law of Conservation of Energy, is also called the **First Law of Thermodynamics** (those are two equivalent names for the same thing).

There's another rule about heat movement that we need to know – and it's much more subtle. When you plonk your coffee down on an iceberg, the coffee cools and the ice warms, never the other way around. The First Law doesn't explain this, necessarily. There's no reason why an iceberg shouldn't cool down, giving up some heat, and make your lukewarm coffee boil – at least, not according to the First Law. As long as the heat gain in one place is perfectly balanced by a heat loss elsewhere, that's absolutely fine. Nor is there any reason why thermal equilibrium can't be reached *either* by coffee cooling or ice warming; as long as we satisfy the First Law, it doesn't matter. The problem is that it doesn't happen. And it doesn't happen because the **Second Law of Thermodynamics** rules it out. In its simplest form, it says that heat always flows from hot to cold, never the other way around (unless there's some kind of outside help). Another way of putting the Second Law is to say that energy tends to spread out and dissipate (although it never actually disappears).

Scientifically, people sum up this idea with the somewhat opaque statement 'entropy tends to a maximum', which simply means that the Universe naturally moves from order to chaos. That doesn't just apply to heat energy. If you drop a wine glass it generally shatters into dozens of pieces; you never see shattered shards jumping back together to form an intact glass – and that's the Second Law in action.

How much heat does a wintry home hold?

In theory and in practice, the Laws of Thermodynamics tell us all we need to know about 'blowing hot and cold' around the home. You need to heat your home because it's colder outside in winter (Second Law). Any heat lost by your home is gained by the surroundings – the ground beneath it and the swirling air all around (First Law). To keep your home at a constant temperature you have to supply just as much energy in the form of electricity, gas or some other fuel as the building loses in waste heat (First Law again). No matter how much we might like it otherwise, houses will never get *spontaneously* hotter in winter, sucking in heat from outdoors (Second Law), although there is a cunning trick we can pull to make them do something similar (using things called **heat pumps**), which is discussed below.

All of this is obvious – so obvious, in fact, that we ignore the implications much of the time. During the winter months newspaper columnists routinely rant about the spiralling cost of energy, the scandal of fuel poverty (people who can't afford to heat their homes) and the 'obscene' profits made by utility companies. It's ironic that the temperatures we experience routinely on Earth are relatively moderate and constant, by absolute standards. The lowest possible temperature, **absolute zero**, is -273°C (-459°F) and has so far proved impossible to attain even in a lab. The chilliest thing anyone can imagine is the inside of a huge black hole, which is about a billionth of a degree warmer than absolute zero[180]. At the opposite end of the scale the hottest thing scientists have so far created is the Large Hadron Collider (LHC), the atom-smashing experiment in Switzerland, where internal temperatures can reach a sizzling five trillion degrees – about 350,000 times hotter than the core of the Sun[181].

Extremes like that make our own battles with hot and cold seem trivial. But whether you're locked in a black hole, spinning in the Large Hadron Collider or shivering in a chilly shack in Antarctica, there's no ignoring the laws of physics. If the outside temperature is (say) zero degrees, and we want a cosy indoor temperature of 18–20°C (64–68°F), the Laws of Thermodynamics explain, unequivocally, that there will be a cost attached to our comfort – and even allow us to work it out. So let's do just that.

Getting warmer

Where do we start with the numbers? In theory, it's easy. You need to list every single object in your home (including the fabric of the place – all the materials it contains) and weigh them. Measure the outside temperature (say, 0°C/32°F) and decide on the inside temperature you'd like (say, 20°C/68°F). For each material, look up a statistic called the **specific heat capacity** (we introduced it in Chapter 2), which tells you how much energy you need to supply to raise the temperature of 1 kg (2¼ lb) of the material by one degree. It's then simple maths to work out the total energy you need to raise the temperature of all these materials by 20°C – and that (thanks to the First Law of Thermodynamics) is how much energy it takes to heat your home.

One of the interesting things about this exercise is grasping the idea that a *thermodynamic* home is much more than just a brick box stuffed with air. If you've ever gone away for a couple of weeks in midwinter and left off your heating entirely, you'll have discovered that it can take two or three days to warm your home up again. Why? It's not simply because it's cooled down much more than usual, but because every single atom or molecule in every single object in your home has (in theory) lost some of its kinetic energy. To heat your home to its usual state of cosiness you have to heat every atom inside it. That's every atom in every chair, table, book, pillow, pen, pencil, picture frame – absolutely everything – and that's why it takes so long. It takes time to pump energy deep inside all these things.

It's difficult to estimate how much heat there is in your home by totting up your stuff, but there's another, more approximate method that can be used. Suppose you live in a modest two-up, two-down terraced home and there are four large electric storage heaters inside it. It's a reasonable assumption that it might take a couple of days' worth of heat to make a house like this fairly warm, which would mean that your four heaters would need to be working flat out all this time. Like everything else in the Universe, storage heaters obey the Law of Conservation of Energy (the First Law of Thermodynamics). Assuming they cool completely from morning to night, the heat energy they pump out during the daytime is equal to the electrical energy they absorb during the night. If the four heaters are charging up for seven hours, for two nights, that makes $4 \times 7 \times 2 = 56$ hours of electricity they're soaking up. If all the heaters are identical, and each is an old-fashioned beast rated at 3,500 watts, that gives 196 kWh (approximately 200 units of electricity), or about 700 megajoules. That, then, is a ballpark estimate of how much heat a small home stores. Compare it with our energy tables in Chapter 2 and you can see that it's roughly equivalent to burning 5 gallons (23 litres) of petrol.

Do you really need to heat a home at all? The Law of Conservation of Energy tells us that most of the energy in the food we consume eventually reappears as heat that we give off to our surroundings. So, with enough people in your home, you could heat it without any radiators at all. How many people would you need? Sitting at a desk or walking around, you radiate about 100–200 watts of heat – roughly as much as one or two large incandescent lamps[182]. To heat a room as effectively as one large storage heater would, you'd need about 35 people sitting down (or 18 pacing around, wearing out your carpets). To replace all four storage heaters you'd need to be able to seat 140 people (or have 70 people wandering around). That's why concert halls need powerful air conditioners.

Getting cooler

Cooling your home can be far more tricky than keeping it warm. Back in the mid-19th century, long before fridges and air conditioners were invented, sweltering Londoners had to rely on shipments of ice blocks imported from places such as Norway. The enterprising Carlo Gatti shipped up to about 400 tonnes across the North Sea at a time[183].

Why is cooling a home so hard? In theory, heating and cooling are exact opposites, so cooling your home should be no more difficult than heating it. If you have a cup of cold water at 10°C (50°F) and you want to heat it by 90°C (162°F) to make it boil, you have to supply a certain amount of energy; you get the same amount of energy back from the water when it cools from 100°C (212°F) to 10°C (50°F). That's the First Law of Thermodynamics in action – from two different directions. However, this doesn't mean heating and cooling are reversible mirror-image processes: you can't flip an electric fire into reverse so that it soaks up heat from a room and cools it down. Nor can you scoop the heat from a room and stuff it back inside a lump of coal ready for reuse tomorrow. Why not?

Heat energy flows from hot to cold things by three processes, called conduction, convection and radiation. In **conduction** hot things touch colder ones and pass heat across by direct contact. The energetic molecules in the hot object directly transmit some of their energy to those in their cooler neighbours. **Convection** carries heat through snaky swirls of rising and falling gas (or liquid). For example, when you heat a saucepan full of soup, the liquid at the bottom, nearest to the flame, warms, becomes less dense and sneaks upwards, pushing cool soup out of the way so that it tumbles back down. The pattern of rising, warm soup and falling, cool soup slowly cycles heat energy through the pan. **Radiation**, the last of the three processes, involves beaming heat through air or empty space in invisible rays (you might remember that we encountered it as infrared radiation in Chapter 8). Radiation (which has nothing to do with dangerous atomic radiation) is what makes your cheeks burn when you sit near a campfire. You warm up even though you're not touching the fire and (because you're outdoors) despite the lack of any appreciable convection.

When you flick on a three-bar electric fire in your sitting room, three red-hot twists of metal radiate heat into the room, systematically warming every other object. As each item in the room warms up, it too becomes a mini source of heat, passing energy on to other objects by conduction, convection and radiation. Nevertheless, there's a basic asymmetry between heating and cooling that stops the process working in reverse.

Imagine if you could invent a three-bar electric cooler with three ice-cold, bright blue bars of metal attempting to suck heat from the room. It simply wouldn't work. First, cooling is not the opposite of heating. One hot thing (a fire) can easily radiate heat to many cooler ones (the objects in a room): the Second Law of Thermodynamics tells us that heat energy naturally dissipates and spreads itself out. However, many hot things (the warm objects in a sweltering summer room) can't effectively conduct, convect or radiate their heat back to one, discrete colder object, cool themselves down and warm it up at the same time. Second, even if the bars could suck heat from all the objects in the room, where would it go? There is no 'sink' to remove the heat. A three-bar fire turns electricity into heat, drawing in energy made outside the room through its power cable: it's part of a wider process that transports heat from the outside of the room to the inside. There's no easy way of running the process in reverse. If you dump a large block of ice in your fireplace, it will heat up and melt, taking energy from its surroundings. It cannot, however, remove energy from the room completely, and once it's melted it's of no more use. A similar argument explains why you can't cool your kitchen by opening the fridge door: all the heat 'sucked in' through the front would simply reappear around the back.

Thus there seems to be a fundamental difference between heating and cooling a home that we cannot get around. Why is this?

Cooling versus heating

Heating up something and cooling it down might be equal, but they're certainly not opposite. Heating makes things more chaotic and disordered, while cooling does the opposite, introducing more order and calm. Heating follows the Universe's natural tendency toward chaos; cooling bucks that trend by imposing (artificially) more order. It's easy to warm a room by letting heat energy spread out from an electric fire, but if you want to cool down the same room on a hot day, you'll generally have to take a different tack.

The simplest cooling device, a fan, works by blowing air past hot things, so encouraging them to lose heat by convection or (in the case of humans) **evaporation** (passing air speeds the loss of sweat from your skin, taking heat from your body in the process). That's a mirror image of how a convector heater works, because it simply blows hot air past things and hopes they'll heat up in the process. However, you can't cool a room for very long with a fan alone. If you have a completely sealed, airtight room and you simply switch on an electric fan, all you do is mix up the air and move the heat from one part of the room to another.

Air conditioners (generally used as cooling machines, although they also function as heaters) work in a different way from fans, because they 'soak up' heat from a room and transport it outside. They're quite similar to refrigerators, using pipes filled with coolants (volatile liquids that boil at low temperatures), although they have things in common with fans too, because they suck and blow air. The basic working principle of an air conditioner involves the coolant heating up inside your room (absorbing heat from its surroundings), then getting piped outside, where it cools down (and gives up its heat) before going back around the loop and repeating the process. You might think an air conditioner or a refrigerator violates the Second Law of Thermodynamics, because it's systematically moving heat from something colder to something hotter – in a direction that pokes fun at the laws of physics. What makes this possible is the electricity we pump in, powering an unnatural cycle that renders hot things hotter and cold things cooler, maintaining a temperature difference that would ordinarily (according to the Second Law of Thermodynamics) disappear.

Heating or cooling, it all takes time. Figure out how much heat energy you have to add or subtract in total, and how much you can shift each second, and that tells you how long it's going to take. This is effectively another statement of the First Law of Thermodynamics – and it applies to anything you want to heat or cool, from a house sweltering in summer to a jug of sloshy ingredients you'd like to freeze into something more tasty. Making ice cream usually takes about three hours because it takes that much time to remove heat energy from milky mush (ice cream's ingredients) so it turns solid. There are ways of cheating, however. In 1890 British cook Agnes Marshall developed a cunning quick method of making ice cream when she cottoned on to the idea of using liquid nitrogen (a cooling fluid normally at temperatures of about -196°C (-320°F) for the purpose[184]. Why does it work? Ice cream is mostly milk, which is mostly water, so there are lots of molecules that need cooling down. In a refrigerator it takes a lot of time for all those molecules to lose their energy and freeze. With liquid nitrogen, much colder molecules are used to do the job. Each one can remove more energy when it whistles past your would-be ice cream – so the cooling happens much more rapidly.

GOING UNDERGROUND

When it comes to heating a chilly winter home, burrowing down into the snow-bound ground might be the last thing you'd think of doing – unless, that is, you happen to live in Scandinavia or Switzerland. In places such as these, people make far more use of **ground-source heat pumps**, which haul warmth up from underground and dump it inside their homes. Although that might seem like a violation of the Second Law of Thermodynamics, it's really nothing of the kind. There's plenty of stored heat a few metres underground where the earth is much warmer than the air above. Using a little electricity to stay within the laws of physics, it's possible to drive a fluid down a pipe, pick up some of this heat and pump it high into your home (where the heat is dropped off), before sending it back down for more.

According to the First Law of Thermodynamics, the heat that warms your home is equal to the heat extracted from the ground, but even if you pull up the equivalent of four storage heaters' worth (14 kilowatts, or 14,000 joules per second), you don't have to supply that much energy yourself. Quite remarkably, that means heat pumps are more than 100 per cent efficient. The only energy you need to supply (or pay for) is the relatively small amount that powers the pump and pushes the heat transfer fluid around the circuit. Heat pumps therefore give the delightful impression of creating (heat) energy from thin air – something we know to be

▷

impossible – and, although they're expensive to install, they pay for themselves in around 10–15 years[185].

When summer comes around, you have another trick up your sleeve. Unlike with a conventional heater, you can run a heat pump in reverse to suck heat from your home and dump it back underground. Again, it's cheaper and more environmentally friendly than air conditioning: all you're paying is the price of the pump.

▲ **Heat mining.** *A ground-source heat pump fires a fluid down into the earth to pick up heat and pumps it back to the surface. Inside your home the heat-transport fluid passes through a heat exchanger. The heat is removed and blown into your home by a fan, leaving the fluid cool enough to repeat the cycle. The energy that warms your home comes from the ground; the pump and fan need a minuscule input of energy by comparison. On the face of it, this violates the Second Law of Thermodynamics because we're moving energy from somewhere colder to somewhere hotter. In practice it's fine because we're using a powered pump to make it happen.*

Cheating with heat

Although it might seem contrary to be constantly heating and cooling things inside our homes – snatching ready-made meals from the freezer and microwaving them, for example – there's no real getting around it. It's true that there are other ways to preserve food, such as canning or using preservatives, but supermarkets pander to homes with fridges and freezers: busy modern families tend to cool foods to preserve them.

The amount of energy we use for heating and cooling foods is far from trivial, but it's still relatively insignificant compared with the energy cost of heating and cooling our entire homes. When it comes to keeping houses warm in winter and cool in summer, we can certainly be smarter – using the Laws of Thermodynamics as our guide. It's a given that warm, cosy homes will cool down in winter: the Second Law of Thermodynamics tells us that heat energy must flow from hot things to cold ones. What it doesn't dictate, however, is how fast the heat flows – and that's something well within our control. Due to physical laws you can't stop your warm home from cooling down, but you can slow the cooling just as much as you care to – using effective insulation.

Among the best-adapted winter creatures on the planet, musk oxen have more in common with well-insulated

▼ **Where does your heat go?** *Since hot air rises, you might guess that most of a home's heat escapes through the roof. In fact, three-quarters of it seeps through the walls, floor, doors and windows. If you want to keep a home warm and cosy, you need to prevent heat from escaping, and draughts from entering, in every direction you can think of.*

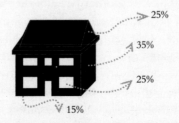

houses than you might think. Shaggy coats that are eight times warmer than sheep's wool help considerably, and explain why native Arctic people call these beasts *oomingmak*, 'animals with beard for skin'. However, the secret is not simply that they're supremely well insulated; they also stay warm and cosy by keeping still and huddling together – just like houses. Standing hibernation, as this is called, lowers their metabolism, making them both 'un-thermo' and 'un-dynamic'.

Holding on to your heat

If heat escapes from our homes by conduction (direct contact), convection (air movements) and radiation (beaming out), the way to reduce heat loss is to slow those three processes down. That's what happens in a vacuum flask. Your hot or cold drink sits in a double-walled bottle with a reflective coating and a vacuum separating its two layers (cheaper versions may use air or insulating foam). Except for the neck and stopper at the top, there's no direct contact between the hot drink and the cold outside, so heat loss by conduction is substantially reduced. The metal container reduces heat loss by radiation, while the vacuum and outer plastic case stop convection.

In theory, insulation in our homes can cheat heat losses almost as well. Insulating materials packed around walls, floors and lofts are generally things like foam, stone wool, vermiculite or plastic with huge amounts of air trapped inside to minimise the heat flow. Double-glazed windows are something of a misnomer because the two thin panes of glass (though insulating) are much less important than the thick, still slice of air sandwiched between them; they'd be better called air-lined windows. Hi-tech, low-E (low-emissivity) windows have a thin reflective coating of a titanium metal compound that bounces sunlight away from your home to keep it cool on summer days. From the opposite side the coating reflects heat produced indoors straight back inside the room to keep it warm and cosy on chill winter nights. (The metal liner in a vacuum flask works in a similar way.)

Most people's idea of a well-insulated home is some UPVC double-glazing and a few inches of stone wool stuffed into the roof – but that barely scratches the surface of what's technically possible. The ultimate insulation would be a house lined with a remarkable lightweight solid called aerogel, nicknamed 'frozen smoke' because that's exactly what it looks like. So effective is it at trapping heat that, according to NASA's aerogel expert Dr Peter Tsou, 'You could take a two- or three-bedroom house, insulate it with aerogel, and you could heat the house with a candle. But eventually the house would become too hot[186].' Unfortunately, although aerogel is 10 times better at insulating than air, it's many times more brittle than glass, so it's likely to be decades before it makes any serious impact on our homes. In the meantime, architects aim for a less ambitious but still quite remarkable improvement in heat performance called the Passivhaus Standard. Developed in Germany in the early 1990s, it's based on the idea of carefully controlling the 'leakiness' of a house so that the only thing you have to heat is the air, which is effectively trapped inside it. The Passivhaus approach cuts home heating costs by around 5–10 times, and reduces annual heating bills on a typical family home to a Scrooge-like £25 (around $38)[187].

Figures like this might sound impressive, but when it comes to insulation, we're still hopeless beginners. You've only to look at musk oxen or snowy sheep to see how much better nature is at beating heat loss in winter. Abundant fossil fuels like coal, gas and oil have made us lazy and complacent, but the rising cost of energy (driven by dwindling supplies, a rising world population and growing fears about climate change) will make us think hard about how we heat and cool our homes in future.

WHY DO COMPUTERS BLOW HOT AND COLD?

Computers – the processors at least – have no moving parts. They're not factory machines or jet engines. They're not bicycle brakes or electric drills. Humans might um and err, and blow hot and cold, but computers leave us in no doubt: put your hand near the 'cooling' fan of a typical PC and you'll probably feel a powerful blast of hot air. Sit a laptop on your knee for more than a few minutes and there's a distinct possibility that your thighs will turn to toast. It's hardly surprising that you feel heat coming out of a computer if you check the temperature inside. My two laptops have built-in thermometers that frequently hit 90–100°C (194–212°F) – quite remarkable, considering that the cooling fan is the only moving part.

Doesn't this strike you as strange? If all computers do is shuffle numbers, why on Earth do they get so hot? The answer boils down (quite literally) to the Laws of Thermodynamics. The power supply on a typical laptop is rated at about 20 volts and 5 amps, which means up to 100 joules of electrical energy is entering the cable every second (the wattage of an appliance is simply the voltage multiplied by the current). The First Law of Thermodynamics tells us that all the energy streaming into the power cable has to end up somewhere – and, save the light beaming from the screen and any sounds that stream from the speakers, virtually all of it is converted into heat. If a hundred joules per second emerge from a computer as heat, that's pretty much the same as the heat given off by a (startlingly inefficient) 100-watt incandescent lamp. That's why your computer gets hot – and why your legs feel so toasty.

Where does all that heat come from? Mostly from electrons struggling through wires like rush-hour commuters hurtling down streets – electrical resistance, in other words. Every piece of wire works like a less dramatic version of a light bulb's filament, heating up a tiny bit as electric current charges through. Modern computers use billions of components, and because most of them are shrunk onto chips the size of postage stamps, the vast heat generated has no easy way to make its escape.

Computers have always been hot stuff – and doubtless always will be. The Harvard Mark I, a classic, part-electrical, part-mechanical computer built in the 1940s, had 800 km (500 miles) of wire stuffed inside, every inch of it generating heat. As we saw in Chapter 12, the even more sophisticated ENIAC, the first proper electronic computer, burned as much electricity as 60 toasters. The famous C-shaped Cray supercomputers from the 1980s were so tightly packed with components that they needed their own built-in refrigerators, which pumped a spooky kind of cooling 'blood' (Fluorinert) through their cases to prevent them from overheating.

Bloggers like to boast that modern computers are different: iPhones and tablets are gleefully green because they need measly power to charge them and have far fewer components[188]. True, but also misleading. First, there are many more of them: there was only one ENIAC, but Apple has sold over half a billion iPhones[189]. Second, mobile devices rely heavily on so-called 'cloud computing' (where your data is stored and processed in giant data centres, connected to you over the Internet, that could be anywhere in the world from Oregon to Bangalore) supplied by Internet giants like Amazon, Apple, Facebook, Google, IBM and Yahoo! According to Greenpeace, the power consumption of cloud computing as a whole increased by a whopping 58 per cent between 2005 and 2010 – and that's particularly worrying because, if it were a country, the cloud would have the fifth biggest electricity demand in the world. Although some of these cloud companies have committed to using renewable energy, several still

▷

source about half their electricity from coal, the dirtiest fuel on the planet[190].

There's no doubt that computers are much more efficient than they used to be, and getting better all the time. Switching from an old desktop computer to a new laptop will cut your energy use by about 50–80 per cent[191]. In the end, there's little room for self-congratulation because there's no escaping the First Law of Thermodynamics: energy has to come from somewhere. If there are more people, using more computers to do more things, there will be a price to pay – somewhere, sometime, by someone.

CHAPTER FOURTEEN

Food Miles

In this chapter, we discover

How much cheese you need to nibble to play a round of golf.
Whether you really can 'go to work on an egg'.
Why idle humans waste more energy than light bulbs.
How the fuel you squirt into your car is much less expensive than the food you pile into your stomach.

If there's any truth in the old saying 'You are what you eat', how come we're not walking, skinny bags of burgers and French fries? The answer is obvious to anyone with a rumbling, grumbling stomach. Our bodies are food factories running in reverse, converting the food we eat into the people we are through a complex, convoluted process called **metabolism**. Now that might sound biological or chemical (biochemical if you prefer) but, like every other activity in the Universe, human metabolism is strictly governed by the laws of *physics*. To put it another way, 'you are what you eat' paraphrases scientifically into 'what you eat is what you can do'. And that's a very casual way of putting the Law of Conservation of Energy (what we referred to in the previous chapter as the First Law of Thermodynamics). Put it even more casually and the banana and chocolate-chip flapjack you gobbled down for lunch translate into half an hour of tennis, a few dozen emails and a great deal of thinking and sleeping.

We tend to think of nutrition in terms of health – biology again – but, like digestion, it's just as much a matter of physics. What you load into your mouth and stomach defines, absolutely, what you can and cannot do in the hours, days and weeks ahead. That's less obvious than it seems, because even the skinniest of us have relatively generous supplies of body fat we can draw on to make up for occasional dips in supply. If you put on weight it's not because you've eaten too much (how much is too much?), but because your body has used less energy than it's been fed – and something must be done with the surplus to satisfy the laws of physics.

It's not only *what* we eat but *how* we eat it that defines who we are. According to some scientists, cooking played a key role in the evolution of big-brained, deep-thinking humans, by making food considerably more nutritious and energy-rich than it would otherwise have been. Cooking itself has evolved dramatically from spit-roasting prehistoric buffalo on an open fire to microwaving ready meals in an office kitchen. As experimental chefs such as Heston Blumenthal readily demonstrate, there's as much science in cooking food as there is in digesting it and converting it into stored or useful energy.

Food versus fuel

Humans need food the way cars need fuel, although that's not the most exact of comparisons. Cars don't grow or repair themselves, don't consume energy when they're sitting idle in the garage and don't put on weight if you fill their tanks too full. Food and fuel both come from the Sun, but by extraordinarily different processes. The oil or diesel you squirt into your tank – petroleum fuel – took a geologically stately *200 million* years or more to form from the rock-crushed, cooked remains of plants and sea animals. By contrast, the tomato you stuffed in your mouth for lunch ballooned and ripened in a matter of *weeks*. A few months ago, the energy it contained (about 35 Calories/150 kJ) was 150 million km (93 million miles) away, high in the Sun[192]. Although your petrol came from exactly the same place, it hasn't seen sunlight since dinosaurs stomped the Earth.

The other obvious difference between food and fuel – two different kinds of stored, chemical, *potential* energy – lies in the way we convert them into mechanical energy that we can use to do things. Cars literally burn energy in sturdy metal 'cooking pots' called cylinders. Petrol turns to wheel-pushing power through **combustion**, the chemical reaction that takes place between something rich in carbon and the oxygen whistling around in the air. Although we cook food and loosely speak of 'burning Calories', no *literal* cooking goes on inside our bodies. When we digest food we convert it into glucose (chemical energy) through the complex process of digestion. Our stomachs and livers convert incoming food into sugars that we can use quickly, or fat that we can store and use at leisure, which is one of the reasons why we don't have to spend our lives crawling on all fours, nibbling at the grass beneath our feet. Although most people use the word respiration as a synonym for breathing, it actually means turning your stored body fuel back into useful energy using oxygen from the air. It's *roughly* like photosynthesis (the light-powered process that turns sunlight into plant food) running in reverse – and analogous to the combustion that happens in your car.

Because of your body's ability to store food for weeks on end, there is no immediate link between what you eat and what you can do. Your body doesn't run out of food the way a car runs out of petrol or a clock suddenly seizes when its spring winds down. The laws of physics tell us that a car can never use more energy than it gets from the petrol you pump into it (although it can get a little help, from time to time, from coasting downhill or being blown along by the wind). In exactly the same way, there's an absolute limit to what you can do and how long you can survive, ultimately determined by the energy content of the food you eat. You might count Calories to limit your weight, but in some sense at least the Calories count you: they set a limit on what you can do.

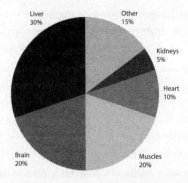

▲ **How your body uses energy.** *Although your brain represents only two per cent of your body weight, it uses almost a fifth of your energy. Your liver is an even bigger energy hog. These figures show your body at rest; during vigorous exercise your muscles would use up to 90 per cent of your energy*[193].

Counting Calories

No one likes burned food, but there's one place where it's quite acceptable. The calorific values listed on food packets are worked out by cremating mini meals in microscopic ovens called calorimeters. In one classic version, called a **bomb calorimeter**, the food is loaded into a tiny metal crucible and burned so that it warms up a water bath all around it. By definition, it takes one ordinary, conventional Calorie (big C) to heat one kilogram of water by one degree Celsius, so if we know how much water there is in the bath and measure how much its temperature changes, we can figure out how much energy the food must have contained (because all the energy from the burning food is absorbed by the water).

Most people find it hard to visualise the energy packed in food; it's unintuitive for three reasons. First, you can't tell just by looking how much energy a certain type of food contains. Is there more energy in a Big Mac or five bananas? They're about the same. Second, as we explored in Chapter 2, most of us have little idea how much energy it takes to do things like climbing stairs or playing tennis, so we've no real idea how much energy – and therefore food – we actually need. Instead of a handy dashboard petrol gauge telling us whether we're empty or full, we have to rely on crude indicators like rumbling hunger and indigestion. Third, the familiar units we use to measure the energy values of food (Calories) are inherently confusing. A food (nutritional) Calorie should be written with a capital C, because 1 Calorie (informal, everyday, nutritional use) is really equivalent to 1,000 calories or 1 kilocalorie (strict, scientific, energy use), which is the same as 4,200 joules or 4.2 kilojoules of energy.

Getting your fill

There's another confusing factor. Just as cars are miserably inefficient, using only about 15 per cent of the fuel you put into them, so the human body is pretty poor at realising the *potential* of its chemical potential energy (food, as we like to call it). Typically, our bodies notch up a mechanical efficiency of about 20–25 per cent, which means if we unlock 100 kJ of energy through respiration, our muscles manage to use only a fraction of it (20–25 kJ or so) to move us around[194]. Where does the rest go? A whopping 60–70 per cent just keeps our bodies ticking over, and the remaining 10 per cent goes in 'administrative overhead', used to process the food we eat. That means we owe Andy, our stair-climbing engineer from Chapter 2, an apology. You can indeed eat a chocolate-chip cookie or two after climbing the Empire State Building because, in reality, your body will burn much more energy as you climb than the bare minimum I calculated would lift your mass against the force of gravity.

Just as inefficient cars can go a long way on a tank of fuel, so our less than perfect bodies can do a surprising amount using energy from the things we eat. Foods make far more sense as fuel if you forget all about Calories and compare joules and kilojoules to start with; you can then directly match the energy in a food to the things you can do with it. Table 4, below, shows some typical foods, their energy value in Calories and kilojoules, and examples of what we could do (in theory) having just eaten them. A slice of bread can power half an hour's reading (you'll need another loaf to finish this book), while chomping down a fat and greasy cheeseburger is enough to power an hour's front crawl. An old advertising slogan used to exhort people to 'go to work on an egg' – and that's perfectly possible as long as you're prepared to walk and it's no further than about 1 km (0.6 miles)[195]. It's also worth noting that fattier foods generally contain the most Calories, which is consistent with the idea that fat has a higher energy density than either protein or carbohydrate.

Even sleeping, thinking and lounging around use energy – a surprising amount, in fact. Remember the musk oxen from the last chapter, with their lazy tactic of surviving the winter through sleeping hibernation? Living things can't stop using energy, but they can reduce it significantly by lowering their metabolism[196]. That's the whole point of hibernating: reducing your energy consumption because there's less energy (less food readily available). In the case of humans, we don't have that option. When we sleep or sit around doing very little, our so-called **basal metabolic rate (BMR)** – our resting rate of energy use – is chewing through about 80 joules per second or 80 watts, which is almost as much as an old-fashioned 100-watt lamp. That might sound horrifically wasteful, but it's nothing compared to hummingbirds, which, famously, have a BMR per unit of body weight that's more than 10 times bigger than a human's[197]. All that buzzing around and hovering makes it imperative for hummingbirds to spend most of their time sucking up energy-rich nectar. Why? So that they can buzz around and hover, and suck up some more. Small animals have to eat lots of energy to maintain the high metabolic

Table 4 **What can you do with food?** *I've attempted to compare the energy values of some typical foods (written in Calories and kilojoules) with the energy you need for some typical everyday activities.*

Food[198]	Calories in food	Kilojoules in food	Enough to power[199]
1 tangerine	35	150	Half an hour of sleeping
1 slice of bread	65	270	Half an hour of reading
½ small can of tuna or 1 large egg	75	315	10–15 minutes of brisk walking
10 large jelly beans	100	420	15 minutes of vigorous aerobics
280 g (10 oz) spaghetti bolognese	250	1,050	45 minutes of housework
1 Snickers bar (60 g/2 oz) or 1 scoop of ice cream	280	1,200	1 hour of gardening
1 bowl of unsweetened muesli (100 g/3 oz), 1 slice of cake, or 1 chocolate milk shake	360	1,510	1 hour of golf
Big Mac beefburger or 5 bananas	490	2,060	1 hour of mountain biking
Cheese steak burger	750	3,150	1 hour of brisk swimming

rate that keeps them warm, which is why a mouse has to eat about 12 per cent of its own body weight of food each day[200]. For a human weighing 75 kg (165 lb), that would be like eating about 9 kg (20 lb) of food a day – or, in weight terms, about 140 Snickers bars.

As an aside, if you think fuel is expensive, looking at the table above gives a completely different perspective. A litre of petrol contains about 34.8 megajoules (34,800 kJ) of energy, and if you buy a lot of the stuff the cost soon mounts up[201]. Compare petrol with fast food, however. A typical burger contains a mere 2,000 *kilojoules* of energy (17 times less), and could easily set you back three times as much as a single litre of petrol. Even electricity is about twice as expensive as petrol[202].

So going on energy content alone, fast food is more than 50 times more expensive than petrol. Is that a fair comparison? It depends what you need to do and how quickly you need to do it. A gallon of petrol could easily propel a car 65 km (40 miles), while a fast-food burger, providing 490 Calories, contains enough energy to power an hour's fast walking at 7 km/h (4.5 mph). If you need to travel 65 km (40 miles), you could do it with less than 5 litres (a gallon or so) of petrol, but if you're walking you'd theoretically need about nine burgers, which would cost about five times more. Then again, you have to buy the car to begin with and pay all the other costs (tax, insurance, maintenance and so on) associated with it.

Fat or lean?

Diets come and diets go – food fads are big business. The classic approach to dieting, counting Calories, makes a lot of sense because it's based on fundamentally solid science. If your body doesn't balance its energy books and ends each day with a surplus of the wrong kind of food, the excess goes straight to your waist, thighs, buttocks and hips. Although obesity is a real and serious public health issue, a little fat here and there is a good thing: it's a prudent, evolutionary defence against lean times. Sheer cultural perversity makes us see fat as horrible biological revenge for self-indulgence. However, it's also our old friend from physics, the Law of Conservation of Energy, in action. Energy has to go somewhere – every joule has to be accounted for. One man's belly fat is another man's potential energy.

The amount you can eat without getting fat obviously varies from person to person – and that's a Conservation of Energy argument as well. Each of us has a different metabolic rate and we all use different amounts of energy during the day. Men and women of different ages have different energy needs. Those needs are greatest when we're about 19–30, young and still active. Athletes and rugby players need to gorge themselves on high-Calorie meals such as steak and eggs; pen pushers need fewer Calories because their bodies have less work to do and need less energy to do it. If you do a sedentary job and you're about 30–50, your daily energy need will be about 2,350 Calories (male) or 1,800 Calories (female). If you have a much more active lifestyle, those numbers zoom up by about 25 per cent to 2,900 Calories (male) and 2,250 (female)[203]. By comparison, polar bears chew their way through about 12,000–16,000 Calories per day[204].

Why do we store fat rather than protein or carbohydrate? Because kilo for kilo and pound for pound, body fat holds about twice as much energy. If your body stores and carries around 0.5 kg (1 lb) of fat, that's twice as much potential energy it has available than if it stored the same weight as protein. You might recognise this 'energy-density' argument from Chapter 5, where we discovered that the world is still chugging around in a billion filthy, fossil-fuelled cars mainly because of the extraordinary ability that fuels like petrol have for storing energy. Body fat is almost as good as petrol and has higher energy density (more per kilo) than other everyday fuels and energy sources (coal, wood, natural gas and batteries).

It's interesting to consider how humankind's total energy needs have changed since the Industrial Revolution. Back in the 18th and 19th centuries, millions of people were driven from back-breaking agricultural labour to slightly less physical work in factories, where coal-fuelled engines and machines were puffing and grinding through the chores instead. These days robotic machines clank their way through the heavy work and people merely poke at plastic keyboards to order spare parts. In the future computers will do most of the brain work as well, and people will probably be reduced (or elevated?) to the status of mere commentators (Tweeters?) on what goes on around them.

If we could calculate the total number of people and machines in each occupation, at each stage of human history, and their combined energy consumption in either food or fuel, what would we find? It's an impossible sum: we don't know exactly what everyone did before the Industrial

▼ **Household expenditure on food.** *Unlike grazing animals, humans don't need to eat all day to get their calorific fill. Even so, food dominates some people's lives more than others. In the United States you'll spend just over six per cent of your money on food; in Pakistan the figure is almost 50 per cent. This chart shows the percentage of household expenditure on food for selected countries for 2012. Drawn using data compiled by US Department of Agriculture Economic Research Service.*[205]

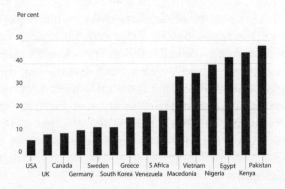

Revolution or how much energy they consumed. It would, however, be interesting to know if we use more energy now (fuelling machines more and our brains and bodies less) than we did back then (when human bodies and brains did everything). Or, if the total amount of *activity* is more or less the same, whether the total amount of *energy* being used – as both food and fuel – is pretty much the same as well. In other words, has all our progress really made us more efficient?

Too many cooks?

Why are there so many cooking programmes on TV? Don't we have better things to do than invent ever-more ways to reheat chopped plants and animals? Maybe. Or maybe our obsession with cooking is fundamentally more important than it seems. That's the conclusion Harvard anthropologist Richard Wrangham reached a few years ago in his book *Catching Fire: How Cooking Made Us Human*[206]. His highly original but essentially simple theory is that the indisputable evolutionary success of humans is down to cooking food, which provides the energy big-brained mammals need more quickly and efficiently, freeing us from hunter-gathering to do much more interesting and productive things – such as cooking and watching cookery programmes.

Anecdotally, that makes obvious sense. Many of the animals we encounter routinely (from cows and sheep to garden birds and bumblebees) seem to spend virtually their whole lives eating or foraging for food – simply so that they can do the same thing again tomorrow. The flocks of sheep I watch from my window spend every waking moment nibbling grass while I manage a quick graze just three times a day, freeing myself to do such important things as watching sheep and thinking about them.

What cooking does is to transform food either so that it instantly has higher energy density (at the most obvious level, a hot drink warms you more than a cold one, reducing your body's need to keep itself warm), or to make it digestible and capable of being metabolised. As Wrangham points out, cooking protein denatures its molecules, so they're easier for us to digest and we get more energy from eating them. That's also why we chew our food. Although chewing requires a certain input of energy, it smashes food into smaller, easier-to-digest particles from which *more energy* can be recovered[207].

The science of cooking

Cooking isn't called domestic science for nothing. You could write a beefy book about the chemical and biological secrets of knocking food into edible shape – and authors such as Harold McGee (*On Food and Cooking: The Science and Lore of the Kitchen*) and Robert Wolke (*What Einstein Told His Cook*)[208] have done just that. Leaving aside the fascinatingly complex biochemical transformations involved in things like baking bread, even the physical aspects of cooking are quite interesting. As we discovered in the previous chapter, the essence of cooking is thermodynamic: shifting heat from the cooker to the food in the quickest and most efficient way possible. Different cooking methods all do it slightly differently.

Oven cooking
From spit-roasting a pig on a prehistoric campfire to rustling up Sunday lunch for the family in a fan oven, basic cooking has changed little in millions of years. Cooking in an oven combines all three forms of heat transfer: conduction, convection and radiation. If there are hot electric bars in your oven, convection makes hot air warm and rise, while conduction passes the heat from the air to your food. Some heat radiates from the red-hot electric elements directly to the food (if they're exposed in the oven); some travels by conduction from the baking tray and cooking shelf. Fan ovens cook faster by accelerating the air to speed up the convection process.

Cooking on a hob

Cooking soup in a pan is convection again, as we saw in the last chapter. The pan itself warms by conduction. It passes its heat to the soup inside initially by conduction, which warms the thick liquid so that it expands and becomes less dense. That makes it float upwards and kicks off the conveyor belt of convection – rising warming soup, falling cooling soup – that eventually warms everything in the pan.

If you don't stir a pan, two things can happen. Some of the soup near the base will stick and burn through conduction. It will also form an insulating layer between the flame, the bottom of the pan and the colder soup above, effectively reducing the convection or stopping it altogether. This explains why you can both burn soup in the bottom of a pan and still have cold, uncooked soup up above it. More interestingly, if you watch saucepan convection at work, you can sometimes see multiple, independently bubbling areas where the soup (or whatever else you're cooking) is rising and falling. Each one of these is a tiny vertical conveyor of rising-warming and falling-cooling soup, technically known as a Rayleigh Bénard convection cell, after its discoverers, Lord Rayleigh and Henri Bénard. If your fancy saucepans have patterns in their bases, that's probably the reason. They seem to promote convection and, in my own experience, make independent convection cells clearly visible when I'm cooking things like rice or pasta with the heat turned up high.

Microwave ovens

The big drawback of a conventional oven is that it has to be blazing hot to heat your dinner. Microwave ovens cook food much more quickly because they beam energy directly into it using short-wavelength, high-frequency radio waves (microwaves). Another advantage of these ovens is that the aluminium box doesn't have to spend half an hour getting hot before the cooking can begin.

How does microwaving work? The energy in the waves makes water molecules vibrate faster inside the food – and since internal vibrations are the same thing as heat, the net effect is internal heating that cooks your food. Although it's sometimes said that microwaves cook quickly from 'the inside out', that's not quite correct. They cook food with more water faster than food with less, so something like a crusty apple pie will indeed be cooked from the inside out simply because the water-rich filling is in the middle. Something that has equal water content throughout (like a lump of meat) will cook more evenly and tend to cook from the outside in, because the microwaves will reach the outside more quickly. As the outside of the meat heats up, it will pass its heat to the inside by conduction, just like in a conventional oven.

Infrared grills and halogen hobs

Grills and halogen hobs that glow red hot beam their heat mainly by radiation, in exactly the same way that a roaring campfire does. That's why eye-level grills can be positioned above pork chops or cheese on toast, and still cook them effectively, even though much of the heat they produce is travelling in the wrong direction (upwards – away from the food) and doing no cooking whatsoever. Halogen hobs (cooktops) are more obviously cooking with 'light' (infrared radiation) than heat. You can tell this because they have an instant heating effect, just like a highly controllable gas flame. Unlike in the case of a traditional electric cooking ring, you don't have to wait several minutes for them to heat up before they can pass their heat on (by conduction). Nevertheless, the glass on a halogen hob gets hot to the touch, so some conduction still takes place.

Induction hobs

Scientifically, these use the most ingenious cooking method of all. Unlike the other types of cooker, induction hobs don't produce heat directly. Instead, they make a sneaky electro-magnetic field that generates ('induces') swirling electric currents, called **eddy currents**, in the metal outer 'skin' of an apparently conventional, iron cooking pan. Unlike normal electric currents, eddy currents have nowhere to go or flow to. Instead, they whirl around the internal crystalline structure of the metal, dissipating their energy by generating heat – so that the pan itself becomes the cooker. The drawback is that you have to use iron-based cookware that will respond to the magnetic field produced by an induction hob; non-magnetic materials, such as aluminium, don't work.

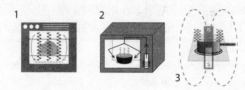

▲ **Three popular cooking methods compared.** *A conventional oven (1) works mainly by convection. A microwave oven (2) beams short-wavelength radio waves from a generator (called a magnetron) into the food, which warms up when water molecules vibrate inside it. An induction hob (3) effectively creates a giant invisible magnet, producing swirling eddy currents inside the pan that dissipate heat and cook the food.*

REACHING FOR THE ENERGY PILL?

If dinner's a drama and tea is tedious, if supper makes you sigh and you find all that chewing a chore, you might be looking forward to the day when you can gulp down a pill and get your entire day's energy needs in one simple swallow.

▷

How feasible is this? A typical A–Z vitamin pill weighs in at about 1.5 g (0.05 oz). If it had to supply our entire daily energy needs (say 2,500 Calories), that's almost 10,500 kilojoules or 10.5 megajoules. We'd need about 666 of these pills to make a whole kilogram and that would give us about 666 × 10.5 or almost 7,000 megajoules of energy.

What kind of stuff could give us that much energy per kilogram? The numbers probably sound meaningless until you flip back to the chart in Chapter 5, where we explored the energy density of different materials. Top of the list was hydrogen, which gives a mere 140 megajoules per kilogram. Even petrol can manage only 50 megajoules. The chances of someone developing a human energy pill 100 times richer in energy than petrol are non-existent[209].

CHAPTER FIFTEEN
Stirring Stuff

In this chapter, we discover

> *Whether there's a right and a wrong way to stir your tea.*
> *Why you can never blow all the dust off a bookshelf.*
> *Why you shouldn't poke cold custard down your sink.*
> *How canal barges help us to understand snoring.*

'When the dust settles' is a particularly ironic cliché – at least in my home. When does dust do anything else and, for that matter, why? When there's air whistling through your windows all the time, why do things ever get dusty or dirty? What about cars? How come cars get filthy when they're brushing through air (and very often rain) all day long? Why doesn't this constant invisible car wash keep your prized machine clean and sparkling?

Curiously, the scientific explanation is all to do with another of my pet domestic peeves: the way people *invariably* make tea and coffee the wrong way. I'm not referring to 'milk first or last?' or other finely nuanced social conventions, but to the way people stir their drinks once the ingredients are safely in the cup. If dust and stirring seem bafflingly unrelated, how about whirling wind turbines, snoring sleepers and cheeky chunks of carrot in perfectly blended soup? It turns out that all these things are finely dependent on how **fluids** – liquids and gases – move through the world around us: they are all intimately connected by exactly the same science.

How not to dust your home
Let's put aside the tricky question of where dust comes from in the first place and consider instead how to get rid of it.

When you come across something really dusty, the natural instinct is to blow the dust away. So you take a deep breath, fill your cheeks like Louis Armstrong – and blow hard. What happens? Some of the dust comes away, but quite a lot of it stays put – for two very different reasons.

First, dust is pretty tiny stuff. You may have read about miners suffering from chronic lung problems such as pneumoconiosis, caused by long exposure to coal dust, or children whose asthma is exacerbated by soot from traffic fumes. The particles of dust involved can be amazingly tiny: the harmful ones go by the name PM10s (particles less than 10 microns, or millionths of a metre, in diameter, which is about 5–10 times slimmer than a typical human hair)[210]. The smaller and lighter something is, the more likely it is to be trapped on a surface by static electricity or the microscopic electrostatic forces that cause adhesion (as we discovered in Chapter 6). Dust sticks because it's small and light, there's a lot of it and it naturally builds up in layers.

The other reason why dust doesn't blow away is much more interesting. Because we're essentially animals, strutting around on our hind legs, we don't notice something very subtle about the weather. You probably won't ever have felt this, but the taller you are, the more likely you are to be blown about by the wind (it's not a big effect, but it's real nevertheless). Although it's obvious that the speed of the wind increases with distance from the ground, especially if you go walking up hills, it's not apparent that, at ground level, the wind speed is zero[211]. By ground level I mean literally within a few atoms of Earth's surface. Fingers of green grass might flutter in the breeze but, at least in scientific *theory*, there's no air movement on the ground whatsoever – even in a force 10 gale. The logical opposite of this explains why wind turbines have to be so tall. If you double the height of the rotor blades, you increase the power they can make by about a third[212].

When you blow parallel to something like a dusty bookshelf, the mini-wind that whooshes from your mouth

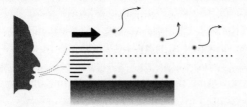

▲ **Why you can't blow all the dust off a bookshelf.** *When you blow, the air above the shelf slides along separate layers, each moving slightly faster than the ones below. Above a certain distance from the shelf, known as the boundary layer (dotted line), the air speed is constant. Near the shelf the air is moving too slowly to shift the dust (grey circles). Slightly higher up, but still within the boundary layer, it's going just fast enough for you to blow the 'dust off the dust'. Even so, some dust always stays behind when you blow.*

has a certain speed. But, logically, at the very edge of the shelf, at a distance of say one atom from the wooden surface, the air has no speed at all. The speed increases gradually as you move away from the surface, over a certain distance that's technically known as the **boundary layer**. Beyond the boundary layer the speed stays constant. You can't blow dust off a bookshelf because the dust is at the extreme lower limit of the boundary layer where the air isn't moving at all. The harder you blow, the more likely you are to shift the dust. Generally, however, you just blow the 'dust off the dust', always leaving some behind.

What holds for dusty bookshelves applies just as much to cooling fans. Have you ever noticed how dirty the blades of a fan get even though it's whipping and chopping, hundreds of times per minute, through the air? The explanation for this is exactly the same: the air right next to each blade is not moving at all. Not only that, but the constant rubbing of plastic against air creates static electricity that makes the dust cling even more. Moreover, what goes for bookshelves and fans goes just as much for cars. Speeding a car through the air involves much the same air flow as blowing on a bookshelf. Although the wind whistles past as fast as the needle indicates on your speedometer, right up close to the metal skin of your car the speed of the air is a big, fat zero. That's why dust and dead flies stick there – inside the boundary layer – until you use a dampened sponge to scrub them away.

GOING WITH THE FLOW

Water and air seem very different, but much the same science applies to both when they move at speed. We lump liquids and gases together and call them **fluids** – because, unlike solids, they can flow – and we name the science of how they move **fluid dynamics**. Aerodynamics is the branch of fluid dynamics that deals with air movements and, in practice, makes vehicles speed along with minimal **drag** (air resistance). One of the interesting things about fluids such as air is that they tend to flow in one of two ways, loosely either 'smooth' or 'rough'.

Smooth flow

As you race down the highway in your bank-busting Ferrari, air slides smoothly around the outer surface, following the sleekly curved profile in lines or streams, helpfully known as **streamlines**. When we say a car is 'streamlined', we really mean that it's been curved and moulded in such a clever way that it barely ruffles the oncoming air. An ideally aerodynamic car disturbs the streamlines as little as possible so that the air flow emerging from the back of the vehicle doesn't look all that different from the air hitting it at the front. When air flows in this way, it's as though it's sliding along in roughly parallel lines. We call this **laminar flow** (laminar means consisting of 'sheets', the way laminated floor is made from thin, parallel sheets stuck on top of one another).

▲ **Laminar air flow around a streamlined car.** *This is the air flow around a reasonably aerodynamic car, much as you would see in a wind-tunnel test. Note how the streamlines bunch up to pass around the car, but remain undisturbed beyond the boundary layer above it. Despite the designer's best intentions, there is always some disruption to the air flow – hence the stray swirls of turbulence behind the car at the bottom.*

Sports cars improve their handling by cleverly redirecting the air that flows past. Many use air dams (air 'scoops' mounted under the front, which reduce unhelpful air flow under the vehicle) and spoilers (mini wings mounted at the very back, which smooth turbulence trailing behind). Formula 1 cars are packed with devices like this to produce downforce that clamps them to the track as they scream around corners at terrifying speeds. At more than about 240 km/h (150 mph), they generate twice as much downforce as their own weight, which is why you may hear people say that you could drive cars like this upside down on the ceiling. The drawback of add-ons like spoilers is that they increase drag and reduce speed. Thus you can't have your aerodynamic cake *and* eat it, although variable spoilers (ones you can extend or tilt to different angles as you drive) give the best of both worlds: high downforce when you need it and low drag when you don't.

Smooth, laminar flow isn't confined to cars – or even air. One of the best places to witness it is on a gently sloping beach. After waves have broken and the water surges up the shore, it's quite easy to see separate sheets of water sliding past one another. You might have a layer on the bottom,

▷

next to the sand, creeping slowly down the beach to the sea, while one or two independent layers of water, above it, are sliding in the opposite direction, from the sea to the beach, at quite different speeds. This looks very much like laminar flow: the sheets of water slide past one another like shiny, lubricated pieces of glass, without any obvious signs of interacting or mixing. Slow rivers also show laminar flow: the flow rate increases gradually from the rocky silt at the bottom, where there may be little or no movement at all, to the surface, where the water flow is fastest.

Rough flow

Of course water and air don't always flow so smoothly. If you're chugging along in a truck, and it doesn't have one of those gently sloping, air-deflecting fairings on the front, you're not going to do the streamlines any favours. As the square front of the truck smashes into the air, it stops or slows some of the parallel-flowing streamlines, while others, beyond the truck's influence, shoot straight past. The result is a swirling stew of mixed-up air that we call **turbulence**. It's chaotically disordered and creates a great deal of resistance to whatever's trying to punch its way through. The turbulent air gets its energy from your truck and that's why it slows you down: drag literally *steals* your energy.

If you're a keen swimmer and your chosen stroke is the crawl, you'll know that the best technique is to make your body as long and flat as possible so that it barely disturbs the water flow. If your head's too high and your back isn't flat, you'll make turbulence in the water around you, slowing your body down, stealing your energy and tiring you out. Swimming in open water usually takes much more effort than swimming in a pool because wind and groundswell, added to currents, make the fluid flow more turbulent: in my own experience, how you hold your body matters less if you're swimming in a particularly rough sea[213].

▲ **Turbulent air flow around a boxy truck.** *The air does its best to wiggle around the vehicle, as before, but some smashes off the square front and bounces straight back, and each disturbed layer of air ruffles other layers nearby, causing considerable turbulent mixing. It takes energy to disrupt air like this. The more the air is churned up, the more energy the truck wastes fighting through it – although that's not inevitable. On a large truck an aerodynamic, roof-mounted fairing will cut drag by about a quarter at top speeds (even though it adds weight). On an articulated truck fairings mounted in front of the trailer wheels will cut fuel consumption by a further 10 per cent* [214].

Putting turbulence to work at home

The idea that fluids move by laminar or turbulent flow (or some mixture of the two) helps us to explain all kinds of things. Going back to our dusty bookshelf, you can see that what's happening when you blow air across a shelf is laminar flow, very similar to what you get in a smoothly flowing river. Each layer of moving air slides slightly faster over the layer beneath, leaving the air at the bottom (next to the shelf) hardly moving at all. Theoretically, it would be much easier to dust a shelf by blowing if you could make the air more mixed up and turbulent – but how would you do it? Maybe you could repeatedly suck and blow, blow from multiple directions or use some other trick to get the air near the shelf moving faster, such as blowing through a straw. In practice, static cling and the basic force of attraction between tiny dust and the shelf beneath make it far more effective to dust with a brush or cloth.

How not to stir your tea

Now we've explored the subtleties of laminar and turbulent flow, you can probably see that if you're trying to 'stir' your tea or coffee, stirring is the last thing you should be doing. Stirring (whether you're working with tea, coffee, paint or sauce) involves sticking something into a liquid and rotating it. If that's all you do, what you probably end up with is round-and-round laminar flow – and not really any mixing.

John DeMoss and Kevin Cahill of the University of New Mexico have developed an astonishing (if slightly hard to describe) demonstration of this that you can watch on online video sites. One of the scientists stands above a transparent, cylindrical container filled with gloopy corn syrup (also transparent), and injects three blobs of coloured food dye into it at different heights and positions. He turns a stirrer an exact number of times and we watch the coloured blobs spread through the liquid into perfect horizontal stripes as the jar rotates. So far, so good: the dye is simply mixing into the liquid. Or is it? Next, he pauses a moment, then reverses the process, turning the stirrer the same number of times backwards. By magic, the dye *unmixes* from the liquid and returns exactly to the three original blobs that were there to begin with.

This almost unbelievable demonstration proves that perfect laminar flow does not produce mixing, even when a fluid goes around in circles; that's why it's completely reversible. When you're stirring tea or coffee all you're really doing is propelling the fluid in circles, in a more or less laminar fashion. Although it's a reasonable way of stirring, it's much quicker and more effective to stir in a turbulent way – so, for example, stirring first one way, then stopping abruptly and stirring the opposite way, and repeating that process. As a rule, I now stir so turbulently that I have to mop up at least 50 per cent of every drink I make from my kitchen's splashed worktops. (Guests frequently accuse me of serving short measures.)

If you make your tea the old-fashioned way, with leaves instead of bags, you might have noticed another curious scientific effect. If you stir vigorously the leaves gather in the centre, at the bottom of the cup, instead of whirling off to the edge as you'd expect. Initially they do move outwards, but, as you stop stirring, friction between the spinning tea and the bottom and the outside wall of the cup slows the liquid there. This sets up a double vortex that cycles fluid from the outside of the cup to the inside, sweeping the leaves towards the centre where, because of their weight, they settle and remain as the rest of the liquid churns around them. Who figured out this usefully useless bit of kitchen-sink science? None other than our old friend Albert Einstein, in a little-known paper published in 1926[215].

CUSTARD CHAOS

Once a cup of tea, always a cup of tea. No matter how much you stir – and whether you do it your way (laminar) or mine (turbulent) – the stuff you end up with always looks the same. Fluids that work in this simple way are described as **Newtonian**, because they follow the straight path of physics laid out by Isaac Newton in the 17th century.

Not all liquids work like this, however, as I discovered to my cost one school morning, aged about 12 or 13. I was studying cooking (laughably called 'domestic science', or perhaps 'home economics') and I'd just been taught how to make proper custard for the first time. It had turned out

▷

reasonably well but, for some reason, I'd neglected to stir it and it burned a little bit on the bottom of the pan. I left it to cool down and forgot it altogether until the end of the lesson, when there wasn't time for a last-minute salvage. I couldn't be bothered to carry half a litre of cold, burned custard in my bag until I went home that afternoon so, with nothing else for it, I decided to wash it down the classroom sink. Big mistake! As I scraped and stirred, the cold custard glooped and galumphed. What began in the pan as a reasonably fluid yellow liquid soon started thickening, dramatically. Oh no! I stirred it some more, but that only made things worse. The more anxious I got, the faster I prodded, the thicker the stuff became. Eventually, it turned to solid chunks of rubber and I had to resort to poking it down the plughole, a bit at a time, with the mini barge-pole end of a wooden spoon.

It took me another decade to discover why this had happened – this time in a science lesson. While most liquids (like water and tea) are Newtonian, others are what we'd describe as **non-Newtonian**. Apply a force to them and they don't simply shift or rotate, like stirred tea. Depending on their internal molecular structure, they get either more viscous and gloopy (**shear-thickening fluids**), or more runny and liquid (**shear-thinning fluids**). Custard (and anything with cornflour in it) is shear-thickening; blood, face cream, toothpaste, squirty cream and phlegm are all shear-thinning. You shake ketchup to make it temporarily thinner and more runny so that you can pour it: applying shear (a physicist's technical word for a force that makes a material shift its shape), you make it thinner. You've probably noticed (and yet never quite registered) that toothpaste works in exactly the same way. It oozes from the tube like a thick and lazy worm but, within a second or two of brushing, has magically transformed itself into a slick runny liquid. Shear-thinning liquids are much less trouble than shear-thickening ones – and always fun when you spot them. Coughing, for example, applies enough pressure to the phlegm in your lungs for it to turn from a solid to a liquid, which can be spat easily from your mouth. Highly distasteful, but highly true.

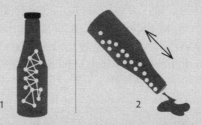

▲ Ketchup: A shear-thinning, non-Newtonian fluid. *Left to their own devices, the crushed tomato fibres inside ketchup lock together in a kind of semi-rigid scaffolding (1). In this form, ketchup is hard to shift. Shake it up and you smash the scaffolding, freeing the stuff to flow smoothly (2).*

Blending in

Laminar and turbulent flow is also hard at work inside kitchen blenders. Most of these vicious machines are *frighteningly* powerful, so they make light work of chopping vegetables into purées and soups. My hand blender is powered by an electric motor rated at 700 watts, almost as hefty as the 860-watt motor in my much bigger and heavier washing machine. Most of us have little sense of what 700 watts really means, but if you look back at the table in Chapter 2 you'll find that it's about 70 times the power you can make (with some difficulty) using a hand crank. Even so, with either hand blenders or the bigger, upright, jug-type blenders, you'll sometimes find that they stubbornly refuse to chop something. Instead of cutting stuff as they're supposed to, all they do is push awkward chunks of carrot round and round in circles. Why? Because once they start rotating, the liquid inside quickly spins at the same speed, so all the blender does is stir by perpetuating laminar flow. If you want a blender to chop things more quickly and effectively, it's often better to pulse the motor. Pulsing creates a more turbulent, chaotic flow in which those stubborn carrots are more likely to drop back down onto the blades than whiz endlessly around the walls like orange motorbikes on a wall of death.

If you're *extremely careful* and watch what you're doing, it's possible to demonstrate some very neat fluid dynamics with a hand blender. Half fill a tall glass jug, or a transparent plastic beaker especially designed for blending, with water, then blend it – just the water. Taking great care (and I can't emphasise that enough – these machines are highly dangerous), slowly lift the blender upwards and watch what happens as you pulse the motor on and off. You'll see amazing swirling flows of fluid (known as vortices) and giant bubbles forming, spinning and dying. If the blender is super-powerful, you'll probably find that it creates enough of an upwards, impeller (water-sucking) effect to hold a heavy beaker full of water against its own weight. As you whiz the blender round, lift it upwards very slowly and you'll probably find that the beaker lifts up with it.

Going with the flow

Have you ever looked closely at the water streaming down from a tap (faucet)? Turn the handle so that the water rushes out, then gradually turn it off again so that the flow dwindles. Notice how the water leaving the open end of the tap is flowing in a much wider stream than the water lower down near the sink. Ever wondered why this happens? Water is virtually impossible to squash into a smaller space, so if a certain volume leaves the top of a tap in one second, exactly the same volume must be arriving at the bottom of the flow, where the stream hits the sink. Because of gravity, water accelerates as it falls, so the bottom of the stream is travelling significantly faster than the top. For the sums to work out,

▲ **Continuity in flowing tap water.** *The water at the bottom of the jet is moving faster than that at the top, so it has to assume a thinner diameter. If that weren't the case, there'd be more water arriving in the sink each second than left the tap.*

the bottom of a water stream has to be correspondingly thinner than the top or there'd be more water arriving in the sink than left the tap. Technically this is called the **equation of continuity**, which just means that the volume of water flowing past any given point, in a given time, is always the same.

In much the same way, when a liquid or a gas suddenly moves through a narrow space, it has to speed up. If you push slowly on the end of a syringe, water shoots out in a fine jet from the needle. If you attached a hosepipe to a syringe, you could make a high-powered jet shoot out constantly. That's how pressure washers work (using electric or petrol pumps to keep the water flowing). Syringes and pressure washers don't create water out of thin air: the same amount of water enters the fat end of the hosepipe as leaves the thin end of the jet. It's just that the water has to speed up, correspondingly, so that there's an equal volume arriving and leaving. This also helps us to understand why wind whistles down alleys and in the streets between tall buildings. Effectively, they're acting like syringes, so the air picks up speed significantly as it blows through the narrow passages.

Architects fail to consider this at their peril. Not only can vortices swirl around buildings; if the effect is severe enough they can even blow pedestrians off their feet.

The equation of continuity also helps us to understand the very literal meaning of the old saying 'still waters run deep'. When a fast-flowing, relatively shallow river suddenly runs into a much deeper channel, it has to slow down to maintain the same volume of flow per second. If you imagine that the depth of the river doubles, but the river banks remain the same distance apart, the volume effectively doubles as well. So the water has to flow half as fast in the deeper section. If it flowed at the same rate as it got deeper, the river would effectively be creating water out of thin air.

THE SCIENCE OF SNORING

Like everything else we've explored so far, moving fluids such as air and water have to obey that golden rule of the Universe we know as the Law of Conservation of Energy. Fast-moving fluids have higher speed, so they have more kinetic energy. But nothing can create energy out of thin air. So if a fluid suddenly starts moving faster, gaining energy, it has to lose energy somewhere else to make up for it – and it does that with an immediate drop in pressure. In physics this is called the **Venturi Effect**, and it helps us to explain all kinds of unusual fluid phenomena you might have witnessed, but never really understood. For example, if you're pootling upstream in a canal barge and there's another barge drifting alongside at the same speed, the two boats often bang into one another. That's because the water has to speed up to flow in between them, so it drops in pressure and pulls them together.

Then there's snoring. While you're asleep air whistling through your pharynx (the upper part of your throat, at the back of your neck) speeds up and experiences a drop in pressure. That causes the pipe to close briefly then open again, fluttering between open and closed with the nasty rasp we call snoring[216]. Why do some people snore and others don't? Slovenian ENT doctor Igor Fajdiga attempted to find out by studying 40 patients who snored loudly, moderately or not at all. He discovered that when snorers breathe in, their pharynx narrows more than it does in non-snorers. The more your pharynx narrows and flutters when you breathe in during sleep, the louder you'll snore[217].

▲ **Snoring science.** *Air accelerates as it whistles through your narrow pharynx, making the pipe vibrate and flutter – and causing the sound we know as snoring.*

Fluid matters

Who cares about taps, hosepipes or badly stirred tea? Dusty bookshelves won't prevent the world from turning – and even bad blenders always chop things in the end. Does any of this really matter? Surprisingly, the answer is 'yes'.

If you drive or cycle to work, aerodynamics makes your life easier or harder, costs you money (in petrol) or sweeps you to work faster and with less effort (if you have a good cycling posture and wear a streamlined helmet). What powers your home? Electricity zapping through wires, but probably also gas and water whistling and burping down pipes, subject to the laws of fluid dynamics. Inside your body fluid dynamics keeps you healthy: blood bangs through your arteries and veins in just the same way that white water whips between the banks of a river – and air fills your lungs this way too. So fluid dynamics is much more important than the genteel niceties of dusting and stirring might suggest. Fluids power our lives, and science tells us why.

CHAPTER SIXTEEN
Water, Water

In this chapter, we discover

> *What gurgling drains and fountain pens have in common.*
> *Why a toilet is a good place in which to make a sandwich.*
> *Why water's still the best thing for heating your home.*
> *How science can save you money in the shower.*

There's no escaping water. Babies are big fat bags of it (about 80 per cent water, compared with 60 per cent or so for adults)[218]. Earth's much the same; curiously misnamed, about 75 per cent of our planet is covered in water – and it feels like rather more than that if you're locked in an Indian monsoon or a soggy, miserable, northern-hemisphere winter. Even our homes are dominated by water. You might think that electricity powers your home, but there's water glugging in and out all day long.

I find that one of the best ways to contemplate water is lazing back in the bathtub, watching the colours swimming in soap bubbles[219]. Now there are all sorts of reasons why you might want to linger in your bathroom: obligation, relaxation and restoration could be high on your list. One thing that probably doesn't cross your mind while you're bathing or singing in the shower is science. Putting hygiene to one side, what does science have to do with basic body maintenance? The answer is 'quite a lot'. Body maintenance is powered by water – and water is powered by science.

Why do we like water?
H_2O – the trickling, gurgling power behind every bathroom – is life's defining substance. We fire space probes

up to Mars and parachute robots down to its surface in
search of this alchemical, life-giving liquid: where there's
water, we reason, there's sure to be life. Rolling rivers,
tumbling oceans, sparkling lakes, woolly fog – there's
something so magical about the many faces of water on
Earth that we quite overlook its *chemical* character. Back in
1781 the brilliant English scientist Henry Cavendish carried
out a famous laboratory experiment, burning hydrogen gas
in an air-filled jar (containing oxygen) so that drops of dew
formed on the glass[220]. If all our water were manufactured in
that way, would we still revere it so much? If it were called
dihydrogen oxide, would we order it so happily in a pricey
restaurant?

You can tell how important water is by measuring how
much of the stuff you slosh around. An average American
family splashes, glugs, guzzles and steams its way through
about 1,500 litres (390 gallons or 80 buckets) of the stuff
every single day[221]. You might think that's impossible, but a
typical modern toilet can use 12 litres (3 gallons) per flush,
while a really old one might thunder through twice as
much; even eco-efficient washing machines suck in 35–50
litres (9–13 gallons) on a full load[222]. These two things alone
(the toilet and the washing machine) swallow half your
water between them. A bath gobbles up 100 litres (26 gallons)
and a 5-minute power shower can hiss through just as
much[223]. It really doesn't take four people that long to flush,
wash, bathe and shower their way through 80 buckets.

We just love water. When a waiter carries that jug of
dihydrogen oxide to your table, you rush to fill your glass
and knock it down, for refreshment. What could be purer,
more life-affirming? Yet by contrast, as we explore more
fully in the next chapter, we also use water to clean our
clothes because it's an amazingly effective solvent, which
makes a reasonable job of dissolving all kinds of things. In
practice this means, almost by definition, that there's no such

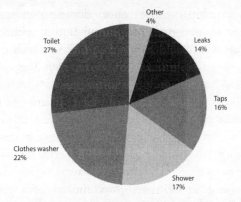

Other
4%

Toilet
27%

Leaks
14%

Taps
16%

Shower
17%

Clothes washer
22%

▲ How we use water in our homes. *The percentage figures used in this chart have been rounded off. (Based on data from American Water Works Association Research Foundation. 'Residential End Users of Water 1999', quoted by US EPA 'Indoor Water Use in the United States').*

thing as 'pure water'. Even tap water, deemed clean, is packed full of dissolved minerals (and a raft of other things we'd prefer not to consider). Earth's water supply is strictly limited and completely recycled: every time you drink you guzzle down molecules that were once peed by Isaac Newton, Albert Einstein and every other favourite scientist from history[224].

We happily scrub our bodies with soap and water for the same reason that we wash our clothes this way, though we could use other things for cleaning instead. We might, for example, borrow the techniques of dry cleaning and shower our bodies with some sort of non-toxic, chemical fluid. Or we could, conceivably, fire some sort of abrasive powder at our skin – a bit like sandblasting the caked black traffic fumes off a medieval cathedral. We could lick ourselves clean, like cats and cows. Why don't we do these things? Because soap and water are inexpensive, abundant and extremely effective. Short of wet wipes (which have the soapy water built into the cloth), I can't recall anyone ever suggesting cleaning the human body in any other way. Although the financial rewards for creating alternative methods could be huge (imagine the market for a dry cloth that got your whole body squeaky clean and meadow fresh with a 30-second wipe before work), it never occurs to inventors to try to introduce them. That's because we take water for granted – it seems perfect in every possible way.

Warming water

If you're going to strip off your clothes in a chilly old bathroom, better clean yourself with something hot and plentiful. As we saw in Chapter 2, water has a huge **specific heat capacity**: it takes more energy (4,200 joules) to warm a kilogram (roughly a litre) of water by one degree than it takes to heat almost any other substance by the same amount. The reason for this is that each molecule of water is built from very light atoms (hydrogen is the lightest of all atoms and oxygen the eighth lightest), so there are more molecules in a kilogram of water than in the same amount of almost anything else you can think of. Each molecule can soak up a certain amount of heat by vibrating or moving about in various ways. So what gives water its heat-absorbing power is the sheer number of its molecules.

And what power it is. To give you some idea, suppose you fill a kettle with a litre of water (roughly 2 pints), then (if such a thing were possible) build a similar kettle as a lump of solid iron. Now suppose you heat both kettles on a stove for the same amount of time so each absorbs the same amount of energy. Eventually the water will boil – but what will happen to the iron? It won't melt, but it will get very hot indeed: its temperature will rise by a frightening 700°C[225]. The fact that it takes a huge amount of energy to raise the temperature of water makes it a perfect vehicle for transporting heat from one place to another, which is why, when you're sitting in the bath thinking about the science of water, it's *water* you can hear gurgling through the radiator on the opposite side of the room, keeping you cosy and warm.

Why water?

There's no automatic reason why central heating systems have to be filled with water. They could just as easily be stuffed full of oil or gas; they could even be solid iron rods carrying heat from room to room by the sheer force of conduction. But that simply wouldn't work as effectively. Imagine a continuous iron bar running in a convoluted loop right through your home, following more or less the path that your central heating pipes take[226]. Now imagine one end of the bar being heated by a gas boiler or a coal fire – it doesn't really matter. For that bar to heat your room it's going to have to be fairly hot. Think about the radiant bars in an electric fire, which are literally glowing red hot, typically at temperatures of about 750°C (1,400°F), and with shiny reflectors wrapped behind them so that they throw their heat into the room[227]. The bar is going to have to stay pretty much that hot as it snakes all the way around your home, through every room, upstairs and down again, and finally back to the boiler.

The Law of Conservation of Energy tells us that if the bar is heating each room in your home, it has to be losing as much energy as it's giving out. So the bar must cool dramatically – and must do so from room to room. There's no way our hypothetical heating bar can simultaneously give out enough heat to warm each room and still retain enough heat to pass to the next room. For that to work the very last room in your home would need the bar still to be hot as it passed through, which means the very first room would have to be correspondingly hotter. The upshot is that the bar simply cannot stay hot enough and extend far enough to heat all the rooms equally, while remaining at a safe enough temperature not to set fire to the house or burn its occupants.

Water holds heat

Does it seem confusing that a red-hot bar can't heat your home? The explanation comes down to the difference between heat energy and temperature. Instinctively we assume that hot things contain a lot of heat energy, which is not always true. As we saw in Chapter 13, there's a basic difference between the temperature of something (how hot it is) and how much heat energy it contains; that's why a chilly iceberg can contain a lot of heat.

The science teacher's favourite example of this is the question of why you burn your mouth on the hot filling of an apple pie but not the pastry crust, even though both may be at similar temperatures. The dry, crumbly crust contains relatively little water, so it has a lower specific heat capacity and contains less heat energy than the fruit filling, which is substantially made of water. When your tongue hits the crust, the pastry cools down and gives off heat to your mouth – but not enough energy to burn you. When the watery pie does the same thing, if it cools down the same number of degrees, it gives off much more energy and that's what burns your mouth. The same reasoning explains why you can, sometimes, briefly hold a pie in an aluminium foil case that's come straight from the oven using just your bare hands. Aluminium has a relatively low specific heat capacity (about five times lower than water's) so, even if it's at a high temperature to begin with, when you touch it and cool it down to your body temperature, it doesn't give off enough heat to burn you.

Getting back to central heating, why does a pipe full of water work so much better than a solid metal rod tracing out the same path through your home? How can the water in a central heating system be dropping off heat in each room and yet still stay hot enough to deliver heat to the next room, the next and the next? It all comes back to water's high specific heat capacity: its amazing ability to retain heat thanks to all the molecules packed inside it. The specific heat capacity of iron is about nine times less than water's. In other words, when a kilogram of iron or steel cools by 10 degrees, it gives up nine times less heat energy than when a litre of water (a kilogram, in other words) cools by the same amount. Although the water in a central heating system progressively cools as it flows out from the boiler, through each room and back again, it can still give out a great deal of heat in the process. And, of course, being a very runny liquid, it can be rapidly pumped back to pick up more heat and repeat the cycle.

That's also why we put hot-water bottles in cold beds instead of, say, heated lumps of solid metal. There was a time when people warmed their beds with copper bedpans, which were like frying pans with hinged lids mounted on long wooden handles that you filled with hot ashes from the fire. Although they were hot, with a high temperature, they

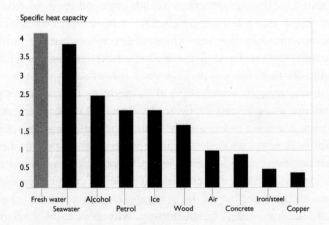

▲ **Why water holds heat well.** *Water holds heat better than almost any other substance: it has a higher specific heat capacity. That means it takes more heat (4 joules) to raise the temperature of a gram of water than a gram of other everyday materials. Generally speaking, metals have low specific heat capacities because they carry (conduct) heat extremely well. The specific heat capacities given here are measured in joules per gram per degree Celsius.*

didn't hold much heat because copper has a relatively low specific heat capacity, and they were inconvenient and quite unsafe. A hot-water bottle in a thick, fleecy cover is still relatively warm the morning after the evening it has been filled, precisely because all those water molecules are so good at holding on to their heat. The same logic applies to your own body. Why does it take you so long to warm up in bed? Because you're mostly a bag of water!

Water in motion

Water's readiness to rush down pipes is just as important as its eagerness to hold heat. If it were a thicker (more viscous) liquid, and trickled instead of flowed, we wouldn't be using it anything like as much as we do. Imagine how long it would take to shower, flush the toilet, do the dishes or wash your clothes if water crawled along at the speed of treacle. Central heating powered by treacle-style water would still work, but much less effectively. The water would cool down in the same way, flowing from room to room, but wouldn't be able to flow back to the boiler quickly enough to pick up more heat.

Plumbing relies on water's willingness to flow, driven by pressure and ultimately powered by gravity; town reservoirs and water towers are invariably built on the tops of hills. Similarly, when you turn on a tap at home, the water gushes out because there's a tank positioned higher up in your home (often in an upper floor or in the attic). Considering it's so low-tech, plumbing is surprisingly effective, but it's not without drawbacks. The sheer speed of water hurtling through pipes can make plumbing extremely noisy. Water is heavy stuff, so when it flows through pipes at speed it has quite a bit of momentum (picture an Olympic bobsleigh team rattling down a run). If you suddenly make it stop by switching off a tap or a valve, the water still moving in the pipe will bounce back, like a wave hitting a sea wall, making the annoying vibrating noise that plumbers call 'water hammer'.

Making water move often involves making air move too – sometimes in the opposite direction. Here's a simple experiment you can try at home (preferably over a sink in a kitchen or bathroom). Take an old plastic bottle of any size, fill it completely with water and screw the cap on tight. Now poke some small holes in the side with a pin and see what happens. Absolutely nothing. Little or no water comes out, because air pressure keeps the holes blocked. Take off the cap, however, and water will stream from the holes. Air rushes in through the neck of the bottle, allowing the water to escape.

Generally, water can only move out of a container if air can get in to take its place. You can see this for yourself by emptying out a water bottle using one of two different methods (use a different bottle, without the holes, if you want to avoid a soaking).

Fill a clear plastic bottle to the neck and leave the cap off. Very quickly invert the bottle so that the neck is pointing straight down, and see how long it takes for the water to empty. While this is happening, watch what's going on in the neck of the bottle. See how there's a kind of violent glugging, with water coming out in intermittent jerks followed by quieter pauses? What's happening is a fight between water and air: air is rushing into the bottle just as water is thumping out. Emptying a bottle of water really means *filling it with air*. You get the glug-glug-glugging because the air and the water have to take it in turns. Some water comes out, some air goes in, some more water comes out – and so on.

Now refill the bottle and try emptying it a slightly different way. You'll find that you can empty the water much more quietly (and often much more quickly) if you very slowly and gently tip the bottle so that it's horizontal. What happens is that the water can stream out from the bottom of the neck while air enters at the top: we get laminar flow with air and water sliding smoothly past one another, in opposite directions, in parallel. There's no turbulent fight between them, and no mixing, so there's no glugging sound.

If you want to empty a washbasin or sink very quietly, you can use the same science to help you. A drain usually gurgles because there's a fight between air and water going on somewhere in the pipe. Most sinks and washbasins have a U-bend in them to prevent smells from coming up, with a small amount of water sitting in the curve. You get that dramatic gurgle at the very end of a washbasin emptying because water, whooshing down the pipe, creates a partial vacuum behind it, which rattles the water in the U-bend. If you have a full basin and swirl the water around as it's emptying, you can create a vortex around the plughole, with water whirling down around the edges and an air gap in the middle that kills any gurgling. Keep swirling and you can often make the water go down without any sound whatsoever. Try it for yourself.

Writing with water

Now this is all very interesting, but does it have any practical use? Of course. If you're a bit old-fashioned, like me, and still use a fountain pen to scribble your letters and birthday cards, you'll find exactly the same science at work, streaming ink onto the paper (when you want it to) and holding it safely inside the pen (when you don't).

A fountain pen is really just a small bottle of coloured water without a cap on. Turn a pen so that the nib points straight down and all the ink should puddle out on the page – exactly like inverting that bottle full of water. It doesn't happen because the nib is connected to an extremely thin tube linked to the ink cartridge, and air pressure pushing up prevents the inky water from flowing down. It's much the same situation as the bottle with a cap on it and holes pierced in its side. In that case, how does ink ever get out to make a mark on the page? That's a little bit like tilting a water bottle horizontally. The thin tube between the nib and cartridge is cleverly

▲ **How fountain pens work.** *Fountain pens are powered by similar science to plumbing: when air flows in through the 'blowhole' in the nib (black arrow), ink can flow out through the reservoir underneath (grey arrow). Capillary action helps to pull the ink through the pen as you stroke it over the page.*

designed with three ink channels at the bottom and an air channel directly above them. The ink flows down and out of the pen when air can flow past it, in the opposite direction, to take its place. That process is helped by **capillary action**. When you drag the pen on the paper, the paper fibres stick to the ink, pulling it from the pen in a stream, each molecule of ink hauling on the next one as though they were a long, flexible but completely connected chain. Ultimately, however, the ink can get out only because the air can get in.

Toilet tips
A lot of home plumbing relies on the same trick: air getting in so that water can get out. Toilets, for example, only flush so quickly because their cisterns (the water reservoirs at the top) are partly open to the air. If that weren't the case, flushing a toilet would be like poking holes in a bottle with the cap screwed on: little or no water would escape. Toilet waste pipes need air vents at the top (or things called air-admittance valves) to work properly. When you flush, air enters the pipe, equalising the pressure caused by the waste water draining away. Plumbing is all about understanding the power of water pressure and air pressure – and making sure that the two work hand in hand.

Toilets obviously flush when water rushes into the bowl: that's what flush means. But this isn't the whole story. Suppose you take a bucketful of water and empty it, a teaspoon at a time, into an unpleasantly 'full' toilet. You might be disappointed to find that's not enough to empty the bowl, but it's no surprise really. If you have a leaky toilet valve with the cistern constantly dribbling into the bowl, that won't be enough to flush it either.

If water alone is not enough to flush a toilet, what's the secret? Toilets rely on an extra trick: the power of the **siphon**. A toilet flushes for the same reason that a washbasin gurgles when it empties. The basic design of a flush toilet relies on a kinked pipe shaped like a U or an S so a bit of water is permanently trapped at the bottom, making a hygienic seal in the pipe that stops foul smells (and germs) from storming your bathroom. When you flush, about 12 litres (3 gallons) of water thunder into the bowl from the cistern up above. Every drop of water that enters the top of the bowl forces a drop around the U-bend, so another drop leaves the pipe on the other side. But remember the equation of continuity. When a wide volume of water funnels into the bowl, the water that emerges from the other side of the pipe has to flow much more quickly. As the water leaves, it speeds up and pulls more water behind it, creating a powerful siphon – a low-pressure, partial-vacuum, sucking action – that very effectively empties the bowl.

Plumbers sometimes need to empty all the water from a toilet for maintenance – and there's an old trick they use to do it. If you hurl a bucketful of water into a toilet, at just the right angle and speed, you can create a powerful enough siphon that will suck the bowl completely dry. It doesn't always work on newer, low-flush toilets with narrow outlet pipes, and I don't advise trying it yourself unless your toilet is completely clean and you're wearing clothes you don't mind getting wet. Do it wrong and half a bucket of less than fragrant eau de toilette will splash back all over you.

Toilets might seem disgusting places at the best of times, but all that cleansing water, the U-bend germ trap and reasonably regular cleaning makes them surprisingly hygienic; your office is probably far dirtier. Arizona microbiologist Dr Charles Gerba has gone so far as to suggest that you're safer making a sandwich on top of your toilet than on your desk, because while a postage stamp–sized area of your toilet seat contains a measly 49 microbes, the same area of your desk harbours about 10 million[228].

Bath or shower?

If you're an ecofriendly energy watcher, or just worried about your gas and electricity bills, you might wonder whether you're better off bathing or leaping into the shower. At first sight it's a trivially easy question: a bath half full of water obviously takes more energy (and money) to heat than a bath quarter filled, which is roughly how full it will get if you shower standing in the tub with the plug in for five minutes. According to frugal environmental auditor Nicola Terry, a typical bath uses four times as much energy as a typical shower or twice as much energy as a power shower[229]. Whichever way you look at it, showering is always better than bathing, right?

Maybe, but it's worth thinking about the question a little bit more. Why is bathing so inefficient? If you live in somewhere like the UK or the East Coast of North America, and you want to swim in the sea outside the summer, you'll need a wetsuit, plus boots and gloves. This is because water, being more dense than air, robs heat from your body about 25 times more quickly than air at the same temperature[230]. Because of the high specific heat capacity of water it takes a relatively long time to cool down but, as it does so, it can remove a relatively large amount of heat from your body. That's why bathwater needs to be so hot: if it's not heating your body and making you sweat profusely, it's cooling your body core and making your teeth chatter. You can safely take a freezing cold shower for several minutes because there's relatively little water coming into contact with your skin and cooling your body core. But you can't comfortably (or safely) sit in freezing cold water for any length of time without shivering, which is the body's early defence mechanism against dangerous, life-threatening hypothermia.

I've found no good scientific data for the average temperatures at which people take baths or showers, but I would guess (reasonably) that people have the water hotter for a bath than for a shower, because they figure that the water will cool over the 15–30 minutes for which they're bathing and they make it as hot as they can bear to allow for that. My thermostatic shower has a safety switch so that it has to be manually unlocked to go above 38°C (100°F), but I think people make baths a little hotter than this. That suggests a bath is considerably more inefficient than a shower because it's using not just more water than it needs, but hotter water too (and because of the specific heat capacity of water every degree you heat it up by costs you that much more).

If you're worried about the cost of a bath (to your purse or the Earth), showering is surely always better. Or is it? The more powerful your shower, the longer you use it and the hotter it is, the more likely you are to use a comparable amount of energy. Not many people have the time or inclination to bathe every day, but many people shower daily (or even twice daily). Four or more showers a week heat as much water as one bath – and conceivably use more energy, if you're not careful.

Saving energy

What about low-energy, eco-efficient showers? They work by reducing the flow of water through the head, sometimes with a crude plastic washer with small holes punched into it that you insert in the pipe to obstruct the flow. The Law of Conservation of Energy tells us everything we need to know about saving energy in the shower. If you want to use less energy, less total heat energy has to flow through the shower head for the duration of your shower. In other words, you have to use less water or colder water, which means showering for less time at the same water temperature, showering for just as long in colder water or using less water per second for the same amount of time and keeping it at the same temperature. Those are the only options. Whether you skimp on the water flow, the temperature or the time is up to you, but you have to make the cut somewhere.

Even using an ecofriendly shower head reduces the quality of your shower: it must do, by the very laws of physics. A shower cannot save energy unless it cuts down on either water (less flow per second or the same flow for a shorter time) or temperature. Accepting that fact, there's really no real point in spending *a lot of money* on an expensive new shower head to save *a little bit of money* on energy. Simply cut your showering time, turn down the water temperature or shower less often. Think logically and clearly about science – and make it work for you.

THE SKINNY

There are two things you can't escape in a bathroom: water and yourself. Lying in the bath, contemplating your navel, you can simplify the problem a bit further, because skin is about 70–80 per cent water. In the end, water is really all there is.

If humans are bags of water surrounded by skin, skin itself is a bag of water – but we don't notice it because all that water is locked up in cells. After a lengthy laze in the tub, skin takes on a whole new dimension. Lift your hands

▷

clear and your fingers will look as ploughed as a field or as wrinkled as a car tyre. It's a popular myth that wet skin goes wrinkly because it soaks up the bathwater and swells. However, the most recent scientific theory begs to differ. Our fingers actually shrink in water when blood vessels inside them contract. Why? Newcastle University's Tom Smulders has proposed that wet skin has *learned* to wrinkle as an evolutionary response – to give us better grip of wet objects[231].

If skin is mostly water, dry skin is *almost* an oxymoron, but drying is something we all have to do after washing our hands. Getting skin dry in a hurry can be surprisingly difficult, as hand-waving, hurtling commuters often discover in public toilets. Paper towels make a poor fist of it, and even huffing-puffing dryers struggle to do the job to anyone's great satisfaction. Intrepid English inventor James Dyson's Airblade dryers claim to get hands dry in a breathless 10 seconds. They do it using electric motors that race around 90,000 times per minute, sucking air past your hands at 690 km/h (430 mph), which is about 80 per cent the speed of a cruising Jumbo Jet[232].

Wool, metal, fibre glass, plastic – of all the materials you can find on Earth, skin is probably the most impressive: waterproof (effectively), breathable, self-repairing, elastic, attractive, protective against minor impacts and sun damage. There aren't enough superlatives to do skin justice. If anyone ever calls you 'skinny', take it as a very great compliment.

CHAPTER SEVENTEEN

Stain Games

In this chapter, we discover

> *How science makes washing and drying a breeze.*
> *Why sunlight makes clothes look brighter.*
> *Why you can easily dry laundry outdoors, even in the middle of winter.*
> *How microfibre cloths scrub clean, even without soap.*

'You *dirty* thing!' Next time someone says that to you, even in jest, why not take them seriously? Challenge yourself to discover how much truth there is hiding in their words. Did you know, for example, that you excrete about 100 billion bacteria every day?[233] Or that 11 per cent of us carry as many faecal bacteria on our hands as a dirty (yes, I said *dirty*) toilet bowl?[234] Sometimes we make a special effort to scrub up. In October 2011, as part of a publicity stunt organised by Lifebuoy soap to underline how proper hygiene can cut infant mortality in developing countries, 37,809 Nigerian schoolchildren were persuaded to wash their hands simultaneously[235]. Mostly, though, we're nothing like as good at keeping clean as we pretend to be. Some 95 per cent of us claim to wash our hands after we go to the toilet, but only 67 per cent of us really do so (men are much worse than women, with 92 per cent claiming cleanliness and only 58 per cent actually practising it)[236].

Appearances can certainly be deceptive – and that's entirely deliberate. How else do we square these filthy statistics with market research findings that we spend billions each year on laundry detergents (roughly \$4–5 billion in the United States and about £6 billion in Europe)?[237] Prompted by guilt-inducing advertisements for products such as these, the *practice* of cleaning becomes much more important to us than the *state* of cleanliness. As long as there are plenty of detergent bottles crammed under the kitchen sink, it doesn't matter whether we ever use them. We're happy to scrub away at dirty dishes with a sponge, oblivious to the fact that after only one day's use, a 'brand-new' scourer will be harbouring about a billion bacteria; cleaning is all too often a matter of rearranging invisible dirt. Meanwhile, the flip side of a clean home is the filthy planet beyond. There's not much point in keeping your kitchen or bathroom spotless if, in the process, you spew industrial-strength bleaches and detergents into rivers and seas, obliterating fish, wrecking the marine ecosystem and causing toxic pollution that will last for years or decades – domestic filth on a grand and global scale.

Well, never mind. Perhaps our hearts are in the right place. And, as we'll see in this chapter, the science is in the right place too. With a little bit of help from physics and chemistry, we're remarkably good at getting things clean when we can actually be bothered to try.

What is dirt and why is it a problem?

Whatever dirt is, there's a lot of it about. According to esteemed biologist Edward O. Wilson, a mere gram of soil contains about 10 billion bacteria[238]. That's not, in itself, the thing that concerns us. You never hear people saying, 'I really must wash these trousers – they're absolutely full of bacteria!' What bother us are the 'secondary characteristics' of dirt: the sight and the smell of the stuff and, more particularly, what other people will make of it. So we wage war on filth, conscious that we're fighting a never-ending battle on two separate fronts. Clothes get dirty from both directions at once: from the outside (when we dribble ketchup on them, for example, or perch on a dirty park bench), and from the inside too (from our own steamy perspiration and a multitude

of other inconvenient and generally unclean bodily fluids).
Cleaning clothes is mostly about getting rid of 'external' dirt
and 'internal' perspiration.

Why does dirt build up on clothes? Because the things
we wear are designed to keep us warm (a subject we return
to in Chapter 18). Where homes are built from stacks of
bricks, clothes are made from microscopic fibres, twisted,
woven or knitted into yarns and textiles such as wool and
cotton (natural fibres) or polyester and nylon (synthetic
ones). A newborn lamb has a fleece made from an estimated
50 million fibres, so it's reasonable to assume that a typical
lambswool jumper probably contains a good few million[239].
Whether your jumper started life on the back of a sheep or
in a fountain of petroleum gushing from the ground (that's
where synthetic fibres come from), it works the same way
and serves the same purpose: the densely tangled fibres trap
air to keep the heat inside your skin. However, so many
small fibres tangled up so closely soak up dirt. They massively
increase the surface area for things to stick to and, because
fibres are so small, dirt and sweat naturally cling to them
with atomic-scale, adhesive force – just as we explored in
Chapter 6.

Deterring dirt with soaps and detergents

Earth's a world of water – and water is just about the best
cleaner there is. That's largely because molecules of water
are asymmetrical. With two hydrogen atoms stuck to one
oxygen atom in a triangular shape, water molecules are
slightly positively charged at the hydrogen end and slightly
negatively charged at the oxygen end. Like a magnet a water
molecule has two opposite 'poles' – which is why it's often
referred to as a **polar molecule**. And just like a magnet, a
water molecule will stick to things, such as blobs of dirt, and
tug them away. This is the secret that makes water such a
superb cleaner – the 'universal solvent' as it's sometimes
known[240].

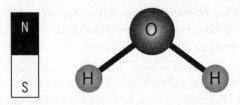

▲ **Water molecules.** *Water, made from two atoms of hydrogen (H_2)
and one of oxygen (O), is a polar molecule. The large oxygen atom is
slightly negatively charged; the small hydrogen atoms are slightly
positively charged. That gives a water molecule two very different ends
and makes it stick to things, including other water molecules, a bit like a
magnet (although using static electricity, not actual magnetism).*

If that's all there was to cleaning, we could manage with
water alone. The only trouble is, water isn't anything like a
universal solvent. It doesn't stick to everything and it doesn't
remove every kind of dirt. The first problem is that water
clings to itself better than to most other things. As we saw
in Chapter 6, that's why it forms droplets, streaks down
windows and forms a skin on the tops of ponds that insects
can dance across. Water's surface tension, as this is called,
has to be broken down before it'll wet things properly.
Generally, polar molecules stick to (and pull away) other
polar molecules, which means that water will readily
dissolve things like salt (another polar molecule). However,
it doesn't work on things like chewing gum, glue, pen or
clothing stains made by organic (carbon-based) chemicals
that have no negative and positive ends to attract polar
molecules.

This is where soaps and detergents come in. Although
soaps *are* detergents, the two words mean slightly different
things. **Soaps** are natural detergents made from sodium and
potassium compounds, while **detergents** are generally
complex cocktails of synthetic compounds, mostly made
from petrochemicals. Soaps work best in soft water but form

an unhelpful, skanky scum in hard water that partly defeats the object of cleaning. Synthetic detergents avoid this problem, but tend to have problems of their own: they can make foam in rivers, for example, or reduce the oxygen content in water, suffocating aquatic and marine life. From now on, I'll refer to soaps and detergents as simply 'detergents'.

How washing gets clothes clean

It's all about teamwork: detergents cooperate with water because neither, by itself, would get your polka-dot socks completely clean. When you load up your washing machine, water trickles through the detergent drawer, flushing the liquid or powder down onto your clothes. The first thing the detergent does is reduce the surface tension of the water so that it wets and sticks to the clothes fibres more readily. Meanwhile the water greatly dilutes the detergent and ensures that there's enough of it to cover pretty much every fibre, in every piece of clothing, in the entire drum full of laundry. While the water molecules would naturally stick to some of the dirt and wash it away, the detergent makes that process a whole lot easier. Detergent molecules surround lumps of dirt and stick to them. As your clothes rumble and tumble around the drum, they're repeatedly bashed and tossed about, helping to break up the dirt into smaller bits and making it easier for the detergent molecules to surround them and do their job. The purpose of the main wash cycle is to ensure that the detergent gets access to as much dirt as possible.

During the rinse cycle more water gurgles into the drum. Now the incoming water molecules stick to the detergent molecules surrounding the dirt and pull them away from the clothes fibres, draining away when the drum empties. Remember that we're cleaning at a molecular level here: tiny bits of dirt are being surrounded by gangs of detergent molecules – and dirt may be trapped deep inside the fibres of your clothes. That's why it takes multiple rinses to get rid of most of the detergent and the dirt it now contains.

Although 'wet cleaning' like this is very effective for everyday clothes, there are two major problems. It's not suitable for cleaning everything because water makes textile fibres swell. That's not a problem if you're machine washing a pair of cotton socks or a drip-dry shirt made from a blend of polyester and cotton, but it's very noticeable if you hand wash a chunky woollen jumper, especially a delicately hand-knitted one – and it's even more of an issue if you've got to clean something like a priceless tapestry hanging on a wall. Unless you're mighty careful, some textiles won't wash and dry in a perfectly reversible way: the fibres don't spring back to their original shape. That's why more delicate textiles generally need dry cleaning, using industrial solvents that cause less distortion of the fibres (although 'dry' cleaning still wets clothes, there's no water involved). The other big problem is that washing leaves you with a pile of dripping wet, horribly heavy clothes that you somehow have to dry.

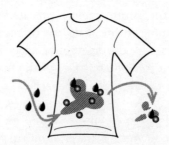

▲ **How soap and water work.** *Soap and water make a pretty impressive, chemical-cleaning team. The soap or detergent (grey blobs) surrounds and sticks to the dirty stain (stripes), breaking it into bits. Water sticks to the detergent and flushes it (and the dirt) away.*

TUG OF DIRT

Mostly we clean with chemistry: soaps and detergents surround dirt and water flushes it away. If you use microfibre cloths, however, which need no detergent, you'll be cleaning in a different way: with physics – and forces.

Ordinary cleaning cloths serve no purpose beyond helping to spread the detergent and wash it away. As the name suggests, microfibre cloths are packed with minuscule fibres something like 10–20 times smaller than those in a conventional cleaning cloth. A decent microfibre cloth will have fibres about 100 times thinner than a typical human hair (just a millionth of a metre across).

How does it clean? The fibres stick to blobs of dirt the way a gecko sticks to the ceiling using masses of tiny 'hairs' on its feet. When large numbers of fibres touch a speck of dirt, they create atomic-scale, attractive electrostatic forces big enough to tug it away from the surface it's clinging to. You don't need any detergent or very much water: the hairy fibres pull dirt away like zillions of small magnets. Indeed, if you use detergent with one of these cloths, you clog up the fibres and impair its ability to clean. Equally, dirt soon clogs the fibres too, so it's essential to wash your microfibre cloths regularly by boiling them up in a saucepan and rinsing them in clean water.

▲ **Cleaning with a microfibre cloth.** *Microscopic fibres adhere to each speck of the dirt and pull it away by sheer force. You need very little water and no detergent.*

The science of drying clothes

Drying clothes is a mundane nuisance, even in the boiling heady height of summer, and even if you have plenty of outdoor space to hang your wet things. But it's an awful lot easier if you think about the science before you start.

How much water in wet laundry?

The first thing we need to appreciate is the scale of the problem. If you've got a washing machine with a detergent drawer, you can experiment for yourself to discover how much water it uses in a typical wash. Just pour bottles of clean water into the drawer until the machine stops filling – and keep count of how much you've used. On the half-load setting I use by habit, my ancient Indesit loads up with about 6–8 litres (1½ gallons) for a main wash, adding maybe another 4 litres (about a gallon) or so at the end to swill the clothes around. Each rinse probably uses almost as much again (6–8 litres 1½ gallons) – and there could be three or four of those before the final spin. On a full load you could be looking at up to 50 litres (11 gallons) of water sloshing through the machine, which is getting on for as much as you drink in a month. That's an amazing amount of water if you multiply it by the number of washes most people do each week, month and year. Assuming a typical family washed a full load twice a week, in one year they'd gush their way through several thousand litres of water, so a couple of streets full of people would use enough water each year to fill an Olympic-sized swimming pool[241].

Now, of course, most of this water comes straight back out again. A typical machine spins at about 1,000 rpm, which is frighteningly fast – much more so than it seems. With an ordinary-sized drum, that translates into a speed of about 85 km/h (50 mph), which is (theoretically) how quickly the drum would race out of your kitchen and hurtle down the street if it ever broke free of its bearings[242]. That's why, if you open up a washing machine, you'll find something like a lump of concrete inside, carefully positioned to counterbalance the electric motor during a high-speed spin. Rotating wet clothes at that speed is fairly good at removing water, but not perfect. If you think about the science of wheels (Chapter 3), unless the drum is lightly loaded and the clothes are all spread around the edge, the clothes that happen to be on the inside are spinning through a shorter distance on each revolution and spinning more slowly – so they don't get anything like as dry as the ones on the outside.

If you weigh a modest load of washing before and after you toss it into the machine, you'll find that it comes out about 2 kg (4–5 lb) heavier than it was when it went in, which means there's about 2 litres (½ gallon) of water still trapped inside the clothes[243]. How about spinning for longer? It's not equally effective every time. Fluffy natural fibres such as wool tend to retain more water than needle-thin synthetic fibres such as nylon. You can almost dry an umbrella with a quick shake or twirl, but you can't do the same with a woollen pullover. The heavier the load you put into a machine, the harder it will be to spin it to top speed and the less water will fly out. Similarly, if the clothes bunch up, so that the drum is unevenly loaded, the machine's electric motor will find it very hard to spin quickly and the clothes will be wetter when they emerge (that's why most machines rock the rinsed laundry back and forth a few times, to loosen and balance it up before the final spin).

Looking at wet laundry as a scientific problem, we can see it slightly differently. What we have to do is separate 2 litres (½ gallon) of water from a mound of densely matted textile fibres – and there are really only two ways to do it.

Drying with force
One way is to use force so that the water comes away from the textiles of its own accord. When you spin-dry clothes you accelerate them in a circle. The washing-machine drum

keeps pushing the clothes round in a circular path, but because of the holes in the drum the water can escape and shoot off in a straight line – into the machine's outer drum and drain pump[244]. Similarly, when you drip-dry, the clothes stay hanging in mid-air, but there's nothing to hold the water molecules there. They trickle down through the fibres, gathering on the hem at the bottom before dripping off to the ground. Drip-drying is really drying using the *force of gravity* to separate water and clothes.

Drying with evaporation

Most of us dry clothes by evaporating off the water they contain: turning liquid water into water vapour. We do that by hanging clothes outside to dry, sticking them in a tumble dryer, draping them on a wooden dryer or even (if we're really desperate and live on the 23rd floor) placing them over a radiator. Because we generally use heat to dry things we automatically assume it's a requirement – but it isn't, by any means. If you hang up a wet shirt to dry, the water trapped in its fibres has to disappear by 'wicking' to the surface (moving by capillary action) and then evaporating away. Liquid water has to turn to water vapour (water as a gas and, in this case, a kind of cold cloud of invisible steam). However, that doesn't automatically require heat.

We tend to think of water morphing between solid, liquid and gas when its temperature changes. So liquid water becomes ice (when it cools) and steam (when it boils). However, you don't always need temperature to make water evaporate. A dish of water sitting out in the open air will evaporate sooner or later, whether it's hot or not. That's why you can remove water from wet clothes even on cold days, but it may take somewhat longer – and happen by a slightly different process. When you boil water in a pan and make steam, you're relying on a steady supply of heat energy to knock the more energetic molecules out of the liquid so that they turn into a vapour: it's the heat that powers evaporation. When you dry clothes in the cold you're relying on passing air to blow water molecules free, so a steady supply of dry wind is the magic ingredient.

Dry is the important word here because the other factor that affects how quickly clothes shed their water (indeed, whether they do so at all) is **humidity** (how much water vapour is already lurking in the air around them). If you happen to live in the middle of a rainforest, there's obviously no point in hanging out your clothes to dry. Similarly, if it's a really hot and humid summer's day, clothes will take longer to dry than if the air is much drier: water is less likely to leave wet clothes and enter a wet atmosphere. By the same logic, if it's a dry day with very low humidity, you can dry clothes outside even if it's extremely cold. Where I live a dry, cold easterly wind often howls inland off the sea giving (ironically) very low humidity. On days like this, even in the middle of winter, it's perfectly possible to get clothes three-quarters dry by hanging them outside for a few hours. The low humidity turns out to be a much more important factor than the lack of heat[245].

A cleaner future?

Washing and drying clothes is a huge waste of time and energy. Although 'wet cleaning' is usually very effective, it's hugely inefficient when you consider the complete process of washing and drying. Wouldn't it be better if we could somehow clean our clothes without water? Will that ever be possible? It might be! The invention of synthetic textiles in the mid-20th century made doing the laundry less of a chore, and there's every likelihood that textiles will improve further.

Few of us relish the idea of wearing an entirely nylon outfit, even if it would wipe clean and dry in a flash. But there must be ways in which technology can improve the clothes we wear. Nanotechnology is already being used to cover textile fibres with stain-resistant coatings that stop

dust and dirt from sticking. You still have to wash these
clothes, but the dirt sticks less readily. What about clothes
made of textiles that dry more quickly – or indeed instantly?
Or clothes we can wash with the laundry equivalent of dry
shampoo (detergent powder you shake over your shirts,
then simply brush or shake away to remove the dirt)? One
company, Xeros, has developed a washing machine that uses
thousands of tiny plastic beads in a wash to reduce water
consumption by 72 per cent, energy by 47 per cent and
detergent by a half[246]. That's the kind of smart science likely
to be cleaning the clothes of tomorrow.

WHAT'S IN THE BOTTLE?

Chemistry and cookery have lots in common. However,
where a recipe can make your mouth water, the chemical
equivalent – the ingredient list on a bottle of detergent – is
more likely to leave you scratching your head or gasping in
alarm. Forget eggs and flour, butter and a pinch of salt; step
forward ionic and anionic surfactants, builders, enzymes,
optical brighteners, solvents, bleaches, dyes and perfumes.
What do all these things do, and why do we need them?

Surfactants, or 'surface active agents', are the main dirt-
busting chemicals in detergents, and there is usually a
handful of different ones in a bottle of detergent. Builders
do a variety of jobs and include such things as zeolite
catalysts, which make life easier for the surfactants by
softening hard water (stealing 'hard' calcium ions from it
and substituting 'soft' sodium ions instead). Enzymes are
chemical accelerators that help the water and detergent to
attack three specific types of natural 'dirt'. The three
common detergent enzymes are proteases, aimed at protein
stains; lipases, which attack fats and grease; and amylases,
which bust apart starch. Of the main, hard-working

ingredients, that just leaves bleaches, which do what
normal bleaches do, and **optical brighteners**. These are
cunning 'marketing chemicals' that give your washing the
notorious 'bluey whiteness', as they used to advertise it in
the 1970s and '80s. Similar to the white phosphor coating
inside fluorescent lamps (see Chapter 10), they convert
ultraviolet light in sunlight into visible light so, in effect,
your white shirt reflects more visible light than is actually
falling onto it.

What of the other ingredients? Modern detergents are
ultra-concentrated and contain so little water that these
chemical ingredients have a job mixing together and staying
in a stable, liquid form. That's why a solvent or two has to be
stirred in so that the other ingredients mix properly and
don't solidify or separate out in the bottle. Dyes and perfumes
don't help with the cleaning, but the marketeers consider
them essential nevertheless. They fool us into thinking that
doing the laundry is a more attractive chore than it is in
reality. The dyes turn the hideous chemical gloop, not unlike
wallpaper paste, into a more attractive, less alien colour such
as turquoise or mauve, while the perfumes persuade us that
we've done our job properly by filling our noses with
fragrance the moment we open the machine.

While detergents get your clothes clean, they don't do
much for the planet. Virtually every ingredient in a laundry
detergent has a well-documented environmental drawback.
For example, surfactants may be directly toxic to aquatic
life[247]; phosphates (sometimes used as builders) reduce
oxygen levels in fresh water, suffocating whatever lives in it;
and solvents can be toxic to both humans and aquatic life.
Detergents can work as endocrine disruptors – the so-called
'gender-bender' chemicals that can cause up to 80 per cent
of fish in rivers to change their sex[248]. Bearing in mind that
aquatic life is part of a much bigger ecosystem, which
includes insects, birds and humans, you can see that the

▷

problem of polluted rivers and seas doesn't stop there: it
rebounds on all of us sooner or later.

The answer? Always read the labels. Think about the
chemistry. Use water, and the mildest, most benign detergents
you can find.

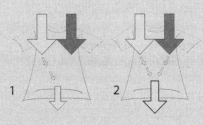

▲ How optical brighteners work in laundry detergents.
*Sunlight contains a mixture of ordinary visible light (grey) and
ultraviolet light (dark grey), which is invisible. When you hold a
white T-shirt in sunlight (1), it reflects some of the ordinary light but
soaks up the ultraviolet light. After washing with a detergent that
contains an optical brightener (2), the fabric surface contains traces of
phosphor-type chemicals (similar to the white coating inside a
fluorescent lamp). These absorb the ultraviolet light and convert it
into visible light. So a shirt washed this way reflects more light
overall — and seems brighter — than it did before washing.*

CHAPTER EIGHTEEN
Dressing to Impress

In this chapter, we discover

> *Why sheep don't shiver in winter.*
> *How a crane pulley can explain the rips in your jeans.*
> *What a wedding dress has in common with a bicycle.*
> *Why a stitch in time (scientifically) does save nine.*

What are clothes but houses you can wander around in? They do pretty much the same job, albeit more colourfully and quite a bit more cheaply. Just like buildings, the things we wear keep us cosy and dry – in some quite surprising ways. Clothes have stolen tricks from nature since the start of human time. Recently, however, textiles scientists have taken to copying plants and animals more directly, with so-called **biomimetic** (biology-mimicking) garments. So we have Olympic swimsuits designed to slash through water like a shark's skin, and porous raincoats that open or close like pine cones. That's just one fascinating direction in which our wardrobes are wandering. In the future clothes are also likely to merge with 'wearable' electronic and medical technology. Party frocks will change colour at the flick of a switch, T-shirts will generate bursts of electricity from sunlight, and cardigans with built-in accelerometers (force meters) will detect when elderly people take a tumble and automatically call an ambulance.

Where a home packs in dozens of hi-tech materials, the textiles in clothes typically use only one or two – yet the similarities are fascinating. The lambswool in your pullover is not very different from the felted, beaten sheep's wool used to waterproof and insulate yurts (Mongolian nomadic homes). On a grander scale, London's O_2 Arena is effectively a giant tent with an umbrella of a roof made from non-stick Teflon (a super-slippy plastic also lurking in your coat and boots). In a modern office block hidden steel girders hold the glass windows high. Stainless steel fibres are probably also hiding in the office carpet, mostly to reduce static shocks, although probably not in any clothes you're wearing[249]. Plastics get everywhere, of course: from the soles of your shoes (polyurethane) and your drip-dry board shorts (nylon), to the frames of your windows (PVC) and the 'glass' in your greenhouse (perspex). The point of all this? That houses and clothes are much the same things, powered by much the same science.

Keeping warm

Most people think of clothes as a kind of heat-proof barrier, insulating by trapping air in their fibres (in something like a thick woollen pullover) and in the air gaps between each layer you're wearing. Wool is a fantastic insulator and – as

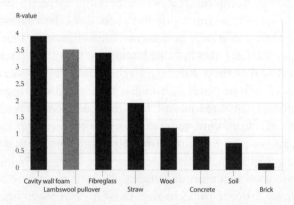

▲ **Wool as an insulator.** *Lambswool is one of the best insulators on Earth – better than almost any conventional building material. This chart shows typical R-values (a measurement of insulating effectiveness) per 2.5 cm (1 in).*

you can see from the chart below – considerably better than most ordinary building materials. However, unless you happen to live in Mongolia, the insulation in your home is likely to be made from cheap, synthetic stone wool cooked from minerals, not sheared from the back of a nibbling sheep.

Now science can easily keep us warm, but it's worth remembering that science also makes us cold in the first place. Deep inside your skin your core body temperature's a cosy 37°C (98°F), but outside in cold temperate countries, the atmosphere is much less than half that, so our internal heat engines are always chugging away at a disadvantage. Making matters worse, the bigger the temperature difference between two things that are in contact, the greater will be the heat flow from the hotter one to the colder one, so on chilly days you lose heat faster than on warm ones[250]. There's no argument about this: the Laws of Thermodynamics force our bodies to cool down, condemning us, in turn, to a life of endless eating just to keep warm.

How do we lose heat? By direct conduction from the core of our body through the outer tissue and our clothes; evaporation from our skin; and convection and radiation from the surface of whatever we happen to be wearing. During exercise about half our body heat is lost through evaporation (heavy sweating), 10 per cent disappears to radiation, and just over a third disappears from conduction and convection[251]. However, it all depends on the weather. On a chilly and windy day, or if you're running or cycling, cold air is constantly brushing past your body, stealing up to half your heat by convection, so a windproof outer layer is your best line of defence. Even if you're wearing multiple layers, it's essential that the outside one is windproof to reduce convectional losses. (If you like sleeping with the windows open, the blankets you throw on a bed do the same job as a windcheater, not simply reducing the rate at which heat fights its way through, but also reducing convection from the top blanket to the draughty air passing by.) On a hot and humid day you're less likely to lose heat by evaporation because the air around you is already saturated, which is why you'll need to fan yourself (increasing convection) to keep cool. On a cold but dry and windless day radiation may account for half your heat loss. Therefore wearing multiple layers to reduce conduction from your body core to the outer surface of your clothing, where the heat radiates outwards, is the best strategy.

Why don't sheep shiver?

Frosty, frozen mornings see you reaching for your warmest pullover – but what exactly is a *warm pullover*? Most likely it's made of wool, possibly merino, a particularly cosy kind of sheep's wool with a very neat, built-in heating effect. Wool is amazingly good at soaking up perspiration (essentially water vapour); it's the most **hygroscopic** (water-absorbing) natural fibre of them all. Merino wool is often used in sports clothing 'base' layers – essentially what people used to call thermal underwear. It has fine, hairy fibres and far more of them, so it produces more heat than regular wool.

Wool is **exothermic** when it meets water: crudely speaking, it converts sweat into heat, in three different ways. First, remember how water molecules are polar, so they cling to things like magnets? They naturally stick to the insides of wool fibres with the hydrogen ends of the water molecules locking onto the cortex of the wool (the cellular structure inside the fibres), making what are called hydrogen bonds. When molecules bond together, they become more stable and give off energy in the process. This is what generates much of the heat that warms your body[252].

The second 'heating' effect is more a lack of cooling – and it comes from the way the water is locked right inside the wool fibres, well away from your skin. Unlike synthetic polyester or nylon, wool doesn't make you all damp and clammy. That's important, because sweating is essentially a cooling mechanism. If you sweat and the perspiration stays on your skin, your body will cool as the water evaporates, whether that's a good idea or not. Since wool completely soaks up sweat, there's little or no perspiration on the surface of your skin, so there's less evaporation and your body doesn't cool as drastically.

There's a third heating effect going on too. As your sweat condenses inside the water-hugging wool fibres, it turns from a gas back to a liquid, giving off hidden heat energy trapped in the water. Where does it come from? When you heat ice to make water or water to make steam, the temperature doesn't rise in a neat and steady way like a straight-line graph. Instead, it climbs smoothly then plateaus for a while, as the water changes state either from solid ice to liquid or from liquid water to steam. The energy taken in when water changes its state is called **latent heat** – and it's used to rearrange the internal molecular structure of the water, loosening solid to liquid (at 0°C/32°F), then liquid to gas (at 100°C/212°F). When you cool water back from steam to liquid or from liquid to ice, you get the latent heat back again.

Together, these three effects explain why wool keeps you warm when you perspire. Understanding this simple science can help you get much better performance from woollen clothes and merino base layers. Suppose you're heading outside on a cold, damp day and – naturally enough – want to keep warm. If you make sure your woollen clothes are completely dry before you put them on, they will be able to soak up more moisture from the outside air, or from your body inside, and give off more heat. So if you're putting yesterday's jumper back on, drape it on a radiator first – not to warm it up, but to dry it out completely before you step out in the damp air.

THE WONDER OF WOOL

Whether you're running your fingers through the woollens in a shop or swiping them through a clothes catalogue on your iPad, science is the last thing on your mind. You'll be looking at size, style, colour, cut – things like that. Now wool might be a cosy textile, but it's still a *material* subject to the laws of physics. This means that we can study and measure it like any other material. So how does it compare to things like plastics and stainless steel?

Wool is made of protein fibres (keratin, also found in human hair and skin), ranging dramatically in length from about 3 to 40 cm (roughly 1 in–1.3 ft), depending on the animal, and packed with cells that are, themselves, made from smaller, stretchier fibres. On the outside wool fibres have a cuticle (made of three separate layers), wrapped in overlapping scales, so that close up they look a bit like the back of an anteater. The scales lock one fibre to another when you press them together, which explains how wool can be converted into felt. The inside of a wool fibre, called the cortex, is made of cells built from thick fibres (macrofibrils), themselves containing many smaller fibres (microfibrils), even smaller fibres (protofibrils) and ultimately a twisted helix of springy protein molecules (the actual keratin).

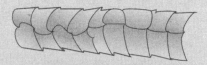

▲ **Structure of wool fibre.** *The 'scales' on the outer cuticle of a wool fibre work like the overlapping tiles on a roof, allowing most water to drain away. However, the cuticle is slightly porous, so wool soaks up water too.*

This Russian-doll arrangement of fibres-inside-fibres-inside-fibres explains why dry wool is so stretchy: the twisted fibres can be pulled straight to about twice their original length, or squashed up to half of it. That's impressively springy when you consider that rubber only pulls to a few times its length before it snaps. Wool might seem pretty weak stuff, but that's largely because we get distracted by its cosiness. Have you ever tried breaking woollen yarn with your bare hands? If you look at the numbers, you'll see why it puts up such a struggle. Its yield strength (the amount of pulling pressure it'll withstand before breaking) is about 10 per cent that of stainless steel[253]. The stretchiness and strength of wool are what make it so resistant to everyday wear and tear.

Wool can resist a bit of water, but not too much. It's hygroscopic (absorbent) because the inner fibres swell outwards like miniature sponges as water penetrates through the cortex. Typically, wool absorbs about 15–25 per cent water, often rather more. Wet wool, especially hot, wet wool, is more plastic than elastic. A pullover washed in warm water increases in size by about 10 per cent, but doesn't immediately spring back to shape. You don't wash woollen pullovers in boiling water because the fibres become more plastic with increasing temperature and deform irreversibly. That's also why freshly washed pullovers are best dried flat. Heat, by itself, doesn't damage wool. The wool dries out, and the fibres start to weaken, become brittle and, as the temperature continues to increase, scorch and char much like hair. Because wool never actually burns when you take a flame away, it's naturally fire resistant – unlike plant-based cotton and flax, which keep on burning when you remove a flame, and synthetic materials such as nylon, which flame dangerously.

Staying dry

Wetsuits (those rubbery swimming costumes favoured by surfer dudes and divers) keep you anything but dry. Even if you wandered around in the rain wearing a wetsuit, you'd still get wet. Very little rain would wriggle through the neoprene, but you'd be so incredibly toasty inside that you'd rapidly perspire: winter wetsuits aren't called 'steamers' for nothing. Indeed, a simple woollen pullover might keep you drier, at least in light rain or drizzle. Wool contains a small amount of lanolin (wool fat), and the fibres are partly water resistant because the scales on the surface knit together and let drips run straight off. Although the yarns in a typical pullover might be fairly widely spaced, a tight-knitted sweater such as a classic fisherman's Guernsey is surprisingly good at keeping out drizzle.

In a pelting downpour wool's no good: you need a heavy-duty waterproof. Most are made from synthetic textiles (essentially plastic fibres) of one kind or another, such as nylon. They don't let the water through because the fibres are microscopically tiny and bonded tightly together. Unlike wool fibres, synthetic fibres themselves don't tend to absorb much water. The water 'beads' on the textile (remains as droplets) rather than soaking in, and it's relatively easy to shake it free, leaving the surface virtually dry. Even so, most of us have had the highly unpleasant experience of wearing a nylon kagoul that's 100 per cent waterproof (on the outside) but drenches you with perspiration (on the inside). Although popular from the mid-1970s to the mid-'80s, cheaper waterproofs like this have now largely been replaced by 'breathable' waterproofs made from materials such as Gore-Tex. Breathable waterproof sounds like a contradiction in terms – and it is. It also sounds like a physical impossibility: how can a piece of material stop water dripping through from the outside but, nevertheless, allow it to seep out from the inside?

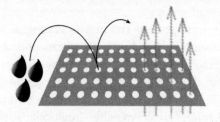

▲ **Waterproof and breathable.** *Gore-Tex contains tiny holes called micropores (white) that let perspiration out without letting rain in. Raindrops (left) are thousands of times bigger than the pores, while the water molecules in sweat (grey arrows) are hundreds of times smaller.*

The explanation is all to do with a basic scientific difference between the water drops in rain and the water vapour in perspiration. A large raindrop is made from a few billion trillion water molecules locked together, whereas in water vapour the molecules are split apart and drifting about as free agents; in other words, a large raindrop is *vastly* bigger than a water molecule. A waterproof and breathable textile such as Gore-Tex takes advantage of the difference. According to the scientists at W. L. Gore and Associates, their wonder material consists of a membrane with microscopic holes about 700 times bigger than water molecules (which simply drift through), but 20,000 times smaller than the smallest raindrops (which, consequently, cannot worm their way in)[254]. That's how the same material can be waterproof *and* breathable at the same time. Inevitably, if you get really hot and sweaty some of the water vapour cools and condenses before it manages to escape, so that droplets of water form inside your clothes and can't steam out. Even with the best breathable textiles, you're still likely to get a little damp inside – and that wetness then starts to rob heat from your body. Despite this, breathable waterproofs are often much warmer than cheaper nylon ones because they keep you much drier inside.

Wearing well

When you were five or six and diving head-first into the school dressing-up box, keeping warm and dry probably wasn't high on your list of priorities. There's more to clothes than boring old survival; safely indoors, feeling comfy and cutting a dash are much more important. Even so, clothes still have to work as dependable, functional materials. They do so in some fascinating ways.

At the beginning of this chapter I compared clothes to houses, and it's quite revealing to explore the similarities a little bit more. Building materials have to balance all kinds of practical considerations, from standing tall and resisting wind to trapping heat and keeping out the rain. We think of houses as structures, even though they are, essentially, 'clothing' for our domestic lives – protection, privacy and tidy family container rolled into one. In a similar way, we can think of clothes as structures that obey the rules of physics just like houses.

In Chapter 1 we saw that a house stays upright by balancing weight on its foundations. The bricks and beams are compressed so that the weight of a house and its contents squashing down are exactly balanced by the atoms under the floorboards pushing back up again. Whether you're wearing an evening dress or a Crombie overcoat, a pair of jeans or a silk kimono, it has weight, just as a house does. However, where a house is supported by compression (squeezing forces), the clothes on your body dangle from your skeleton supported by *tension* (pulling forces).

A heavy wedding dress tugs down on the shoulders in an analogous way to the deck of a suspension bridge dangling from its rope hawsers. The huge, heavy weight of the fabric is shared between tension forces in the seams and individual fibres in a similar way to how the weight of all those trucks and cars is split between the strands of rope in the bridge (or, to make an entirely different connection, how the weight of a bike and its rider hang, effectively, from its hubs by the criss-cross spokes, as we saw in Chapter 4). Think of a pair of jeans as a structure and you'll see them in a totally different way. Note how the main structural member – the waistband – supports the entire weight of the heavy, hanging denim. This creates major stresses at the waist, which is why the tops of good denim jeans are bolted together with industrial-strength rivets.

HOLEY MOLEY – THE SCIENCE OF RIPPED JEANS

I have a fashion-conscious friend who once explained to me how he'd splashed out on a pair of expensive jeans and immediately set to work with a cheese grater, roughening up the denim and introducing racy-looking rips. Most of us reach the same destination after years of wear and tear; he takes a short-cut – quite literally.

It's not too hard to understand why holes appear in the knees of jeans and the elbows of sweaters. When I was younger and more prone to pushing toy cars around the carpet, it was always clear what caused the problem. Years later, with no cars or carpet in sight, it's still the knees that go first. Why? Every time you stand up or sit down, the legs of your trousers shuffle past your knees. Now skin isn't in any sense abrasive, but if you watch what's happening it's easy to see the problem. The fabric strains as it radically changes direction and, if your knee is bent completely, moves more or less at a right angle. What we have here isn't so much a trouser leg as an extremely inefficient pulley.

Pulleys are the ropes and wheels you get on cranes. The rope loops back and forth between a whole series of wheels and, in that way, the weight that's being lifted is shared over multiple rope strands. That's why a crane can lift so much weight, albeit rather slowly. You might have looked up at a tower crane swinging building blocks high across the city and wondered what would happen if the rope broke. But it never

▷

does. The rope doesn't rub back and forth on the pulley, as you might expect, so it never wears out. Instead, the pulley wheel rotates, turning at exactly the same speed as the rope moves, so there's virtually no friction at that point. Much like in the wheels we saw in Chapter 3, the friction in a pulley is almost entirely confined to the point where the wheels turn around their axles, so it does the rope very little harm at all.

Jeans are quite heavy and the top, thigh sections generally work harder than the bottom, calf sections (because they have to support the entire weight of the garment). But the bottoms often rest on your boots and shoes, so the knees are probably the parts that feel the biggest force – and wear the fastest. The trouble with trousers is that knees don't have pulley wheels. Where a pulley allows its rope to slide smoothly with little or no friction, a trouser leg simply rubs back and forth on your rickety bones – so it's only a matter of time before a hole appears.

▲ **Pulleys and knees.** *A pulley doesn't wear the rope that passes around it because the whole wheel rotates at the same speed as the rope to move it up or down. Except for a little rubbing here and there, the only significant friction is between the pulley wheel and its well-greased axle (shaded area). Knees wear trousers because, unlike the rope in a pulley, the knee stays still and the fabric rubs repeatedly back and forth across it.*

A stitch in time

Thinking of the forces that act on clothes helps us to understand why they make us look good or bad. Where buildings are designed to be rigidly dependable, resisting whatever forces Earth and its weather can throw at them, clothes have to work much more subtly and adaptably. They have to be flexible and shift with our bodies, and the forces they feel are constantly changing. But clothes respond to forces in a quite different way from other materials (such as those in buildings).

Take, for example, a typical jumper, knitted from wool or acrylics in precise rows. Look at most garments – from socks and shirts to dresses and coats – and you'll see the same thing: warp and weft patterns that are, very broadly speaking, analogous to the rows and columns of atoms you'll find in something like a metal bar. However, where a metal bar is equally strong in every direction (**isotropic** is the technical word), textiles are stronger in some directions than others (**anisotropic**). They will stretch more on the diagonal than parallel to the warp or the weft, because there's much less resistance in the diagonal direction. That's why dressmakers often rotate fabrics so that the warp and weft hang on the diagonal (a 'bias cut'). As a bias-cut dress drapes it stretches and clings to a woman's body, naturally flattering her curves as she moves, and stresses the fabric in some directions more than others[255].

Clothes obviously wear out faster than buildings – and for quite different reasons. Friction quickly puts paid to the knees of your jeans, but it's not the only force attacking your wardrobe. If you bend a paper clip back and forth, it breaks because the repeated stresses and strains cause cracks to ripple through the crystalline metal structure. A similar thing happens to clothes. If you fold and refold them in exactly the same place, you weaken the fabric at that point and cause something analogous to metal fatigue. Office shirts wear quickly at the collars because every time you tie your tie, you fold the fabric up and down. Eventually the cotton threads snap like paper clips through fatigue – repeated back-and-forth stress in the same place.

Once clothes start to fail, forces finish the job. A tiny bit of wear in the knees of your jeans will rapidly turn into a gaping hole. Why? Because due to the weakness in the fabric the remaining threads have to support more weight than they did before, which makes each of them more likely to fail in turn. A slight rip in your jeans concentrates the stress in exactly the same way as the grooves in a bar of chocolate (Chapter 9) or a crack in the metal fuselage of a plane. The more a rip spreads, the higher the tension on the fibres that are left and the bigger the chance of the rip spreading further. The forces are smallest when the rip itself is smallest, so there's solid science behind the old saying 'a stitch in time saves nine'.

THE SCIENCE OF SHOES

Science gets everywhere – even under your feet. Every shuffling step you take puts physics into play.

First, looking at force, walking involves pushing back against the street to go forwards, so it's a perfect example of Isaac Newton's third law of motion. To walk efficiently your feet have to grip the ground securely. It's easy to understand what's happening if you consider the similarity between a car tyre and the rubbery tread on your shoe. Just as a tyre grips the road and helps to fling a wheel forwards so, as you lower each of your feet to the ground, it grips the surface, making you spring forwards without slipping. The less effectively your shoes grip, the harder it is to go forwards by pushing back. It's tricky to walk both on slippery surfaces, such as ice or a wet floor, and unstable ones, such as shingle beaches. As you push back, your foot simply carries on moving and you either lose balance, or struggle to produce enough forwards force. That's why walking on sand or snow takes two to three times as much energy as walking on a hard surface[256]. If the kind of shoes you're wearing interfere with the walking process, that's also a problem. It's harder to walk in poor-fitting sandals and flip-flops because, even though they grip the ground, they slither around your feet, frustrating the process of push-back, walk-forwards. So the fit of a shoe is critically important. A loose, wobbly shoe might feel more comfortable than a fitted one, but it's harder to walk in because it makes it harder to produce effective forwards force.

You might think walking uses little or no energy because, although you have to lift your feet and legs, you put them back on the ground again a few moments later. Theoretically, then, you get back all the energy you use lifting your limbs against gravity as soon as your feet return to the floor. In practice, however, there's an overall energy loss because you're having to flex and release your muscles – and the complex human gait has built-in inefficiencies as you swing from side to side and switch from one foot to the other[257]. Each step involves repeatedly bending, then straightening your shoe, which wastes energy through a kind of 'rolling resistance' in much the same way as rotating a bicycle or car tyre does (Chapter 4). Most of us realise that long-distance walking or hiking is harder work if you wear heavy boots rather than light footwear, but it's less obvious that *stiff* footwear also saps your energy.

These force and energy effects explain why Olympic sprinters wear spikes, those basic, fabric running shoes with metal points. If you've ever worn a pair, you'll have been amazed at how light and flimsy they are. But, tied tightly, they do their job superbly: due to the lack of weight you squander less energy shifting your shoes, while the spikes give perfect grip, minimising friction and energy losses, and giving a superb amount of forward spring. Ordinary jogging shoes have much more padding than spikes, so they protect your legs and back from impact injuries as you run and jump. However, too much padding adds weight and wastes energy. As your feet push down then lift back up again, the padding is squashed and released, wasting more energy (like a chunky mountain bike tyre) through greater rolling resistance.

All shoes wear out eventually – and usually in one of two places. Sometimes the soles grind into holes, although not as quickly as you might expect. Given the mileage you clock up, the direct frictional wear on a pair of shoes is actually quite small because, just like a car tyre, the soles slide relatively little as you walk. Most shoes fail by cracking either on the uppers, just beneath your toes, or on the soles directly underneath. That's no big surprise. These are the places where the materials – leather, plastic, fabric or rubber – undergo repeated flexing back and forth. So your shoes fail through fatigue, exactly like paper clips bent back and forth. The fact that they can survive millions of cycles of flexing before they finally crack and split (where a paper clip might manage only a dozen) is pretty remarkable.

Notes and References

Here are the notes and references that tally with the superscript numbers in this book's main text. They are effectively a 'condensed' version; for the full referencing, including links to various webpages and other wonders, please see my webpage at www.chriswoodford.com/atoms.html

1. I've gleaned biographical details about Einstein from the eminently readable biography Isaacson, W. (2007), *Einstein: His Life and Universe* (Simon & Schuster, New York). Dropping out of high school is covered on p. 23, his polytechnic entrance failure on p. 25 and his job-seeking struggles on pp. 58-65.
2. Public opinion statistics quoted here are from both the United States and the UK: see my website for links to the references.
3. About ESB, *The Official Site of the Empire State Building*, accessed 29 October 2013 (see website for the link). 340,000 tonnes = 340 million kg, which is equivalent to about 4.5 million people each weighing 75 kg (165 lb). In 2011, Calcutta's population was roughly 4.5 million people, according to Wikipedia.
4. More precisely, it's 412,500 pascals (Pa) or 60 pounds per square inch (psi).
5. It's roughly four million pascals (Pa) or 600 pounds per square inch (psi).
6. Hayes, L. *et al.* (1956). The Latest from Paris: An All-Plastic House. *Popular Mechanics*, August 1956, p. 89.
7. Hay, D. What Does a House Weigh? Some Mental Heavy Lifting. *Seattle Times*, 19 December 2004. See my website for a link.
8. If that makes you wonder why anything moves at all, the answer is because the action and the reaction affect different things. If you fire a gun, the action is what pings the bullet through the sky; the reaction is the recoil that thumps the gun in the opposite direction. The action moves the bullet, the reaction moves the gun.

9. For every common building material, there's a measurement called Young's modulus (E) – roughly, an indication of how stiff or elastic it is. Young's modulus is calculated by dividing the stress on a piece of material (how much force you apply per unit area) by the strain this produces (how much the material stretches compared to its original length). I've assumed a fairly high strength concrete that occupies about a tenth the floor space of my office block. I've also assumed, in effect, that everyone inside the building is standing on the roof, compressing the entire structure equally throughout its height. In reality, the bottom of the building is compressed more than the top.

10. How Tall Can a Lego Tower Get? *BBC News*, 4 December 2012. See my website for a link.

11. Kunreuther, H. *et al.* (2013). Overcoming Decision Biases to Reduce Losses from Natural Catastrophes. In Shafir, E. (ed.), *The Behavioral Foundations of Public Policy*, p. 405. Princeton University Press, Princeton.

12. For a good discussion, see Cool Formula for Calculating Skyscraper Sway, on the Maths Pig Blog, 21 March 2011 at mathspig.wordpress.com/. Also Building stiffness and flexibility: earthquake engineering, at *Architect Javed*, 16 October 2011, at articles.architectjaved.com/. See my website for the full links.

13. Levy, M. and Salvadori, M. 2002. *Why Buildings Fall Down: How Structures Fail*. W. W. Norton, New York.

14. Taipei 101: About Observatory: Servicing Facilities. *Taipei 101*, accessed 29 October 2013. See my website for the full link.

15. A red-hot slap across the face turns kinetic energy (a moving hand) into light, heat and sound energy; pigeon 'applause' sees stored chemical energy (food) turned rapidly into kinetic energy (flight); fizzing pills convert chemical energy into sound (and possibly heat); winking smoke alarms turn electrical energy from their batteries into light; trapped flies are waiting to be converted into chemical energy (digested food), while taut spider webs are examples of elastic potential (stored) energy.

16. Calculation: Potential energy = mgh = 30 kg × 10 m/s/s × 400m = 120,000kJ. Bob uses 260kJ and Andy uses 380kJ.

This is purely a calculation based on shifting a mass upwards by a certain distance. It doesn't take into account the body's inefficiency or any other forms of energy use on the way, such as scuffing your shoes on the floor (frictional losses), waving to people on the ground (mechanical energy), whistling a happy tune (losing energy as sound) or anything else.

17. If Andy weighs 95kg, his climb to the top requires at least $95kg \times 10m/s/s \times 400m = 380,000kJ$, which is equal to 90 Calories. As we'll discover in Chapter 14, things are not quite this simple: any cookie Andy eats will *not* be converted 100 per cent into potential energy. Dr Anthony Viera from Chapel Hill School of Medicine, University of North Carolina, has proposed food labels that show not just Calories but how many minutes worth of walking it would take you to burn them off. See Fast-food Consumers May Eat Less if Label Describes How Long it Takes to Walk off Calories, at the UNC Gillings School of Global Public Health, 21 January 2013 (full link is on my website).

18. Calculation: Energy = (mass) × (specific heat capacity of water) × (temperature change). For 1 litre (1kg) of water, heated from 10°C to boiling point (100°C), that gives $1kg \times 4200J/kg/°C \times 90°C = 378,000$ joules. Assuming a dynamo is about 5 watts (5 joules per second), that gives 75,600 seconds = 21 hours. Bicycle dynamo figure from various models on Amazon.com (Dynosys models are about 5 watts; Busch & Müller are about 6 watts). However, if you ditch the puny dynamo and harness the cyclist's pedalling power much more directly using a decent electricity generator, you could theoretically capture as much as 400 watts and boil your water 80 times faster – because that's the maximum sort of power a racing cyclist is capable of delivering at top speed. That figure comes from the excellent Wilson, D. (2004), *Bicycling Science* (MIT Press, Cambridge, MA).

19. You can read James Joule's own account of his energy experiments in Shamos, M. H. (ed.) (1987), *Great Experiments in Physics* (Dover, New York), pp. 166–183.

20. Once again, I am thinking *only* about potential energy and ignoring everything else.

21. This calculation also measures *only* how much potential energy Andy needs to lift his body up that distance, taking no account of mechanical efficiency (which will multiply the amount accordingly): 95kg × 10m/s/s × 400m = 380,000 joules/1800 seconds = about 210 watts.

22. A fascinating green-energy buff called Piet demonstrated this to me about 20 years ago in his self-sufficient 'mobile office'. His solar-powered laptop was wired to an inkjet printer that took its power entirely from a hand-crank bolted to his desk. To print each sheet of paper, you had to grind the crank around quite a few times – and it felt about as difficult as turning over a car engine the old-fashioned way. I've never forgotten this, the most vivid demonstration of making energy I have ever had.

23. My estimated hamster power is about 0.5 watts and comes from a guesstimate supplied in Fink, D. Science for Kids: Hamster Power, 13 December 2005, at http://en.allexperts.com/q/Science-Kids-3250/Hamster-Power.htm.

24. These are simplified guesstimates that don't take into account energy losses.

25. The amount of energy you need to achieve something (climbing a hill, boiling a kettle or drilling a hole in wood) is always the same, but you generally need more than this amount because whatever you're doing is not 100 per cent **efficient**. Efficiency is the amount of energy it takes to do something in practice compared to the amount it ought to take in theory. If something takes twice as much energy as it should, it's only 50 per cent efficient. If you want to get your body to the top of a mountain, you'll need to use more energy if you go by car than by bike. That's because a car engine is much less efficient than a bike. Going by car will feel easier because the energy is coming not from your body but from the petrol you're burning. But if you had to pay per unit of energy you use, going by car would cost far more. That's partly because a car weighs about 15–20 times more than you do, so you're lifting a load of metal, rubber and glass up the mountain as well. It's also because car engines and transmissions don't convert all the energy locked in the petrol into kinetic energy that moves the vehicle along.

26. To boil water by stirring with a spoon, you'd need to ensure that all the heat you're adding isn't lost just as quickly to the cooler, surrounding air. Theoretically, you could do it if the spoon spun around in some kind of cunningly designed vacuum flask. That would be a modern spin (in more ways than one) on James Joule's original experiment, where he used a falling weight to spin a paddle wheel inside a closed container of water.

27. James Joule, Letter to the editors, *Philosophical Magazine* 27 (1845): 205, quoted in Shamos, M. H. (ed.) (1987), *Great Experiments in Physics* (Dover, New York), p. 169.

28. Gjertsen, D. 'James Prescott Joule', in Wintle, J. (2013), *New Makers of Modern Culture: Volume 1*. Routledge, London, p. 772.

29. Kinetic energy $= \frac{1}{2}mv^2$ where m $=$ mass and v $=$ velocity (or speed, in this case).

30. Since a car's energy is equal to the square of its speed, doubling your speed means you have four times as much energy, while quadrupling your speed gives you sixteen times as much. That's why a car crash at twice the speed is much more than twice as dangerous.

31. Your potential energy $=$ mgh $=$ 75kg $\times$ 10m/s/s $\times$ 10m $=$ 7500J.

32. Assuming you're travelling at 15m/s when you hit the ground, your momentum is 1125kgm/s and the force from an impact lasting between 0.1 seconds is 10 times this number $=$ in the order of thousands of newtons, which is approaching the sort of force you'll get from an alligator bite (according to our table in Chapter 1).

33. Suppose Archimedes balances Earth on a kind of super-sized golf tee, with the entire weight acting down through a pinpoint. Let's say he sticks his tee on the very end of a lever balanced 1m (3ft or so) from the end. If he jumps down on the other side of the lever using his entire body weight (supposing it's about 75kg or 165lb), how long would the lever need to be to balance horizontally like a see-saw? If the weight of the world is 6,000,000,000,000,000,000,000,000,000kg, it's an easy sum. For a see-saw to balance, the weight times the distance has to be the same on each side of the pivot. So the length of the lever would be six trillion trillion divided

by 75 or 80,000,000,000,000,000,000,000 (80 million trillion) kilometres (50 million trillion miles). Sounds fine in theory, but you see the problem? The distance from Earth to the Sun is a mere 150,000,000 kilometres (150 million kilometres or 93 million miles). Lost in space, what would he use for gravity? And what exactly would he stand on? The more you think about it, the more ridiculous it seems — which is probably why it's such a memorable thought experiment.

34. The energy you use is given by the formula: potential energy = mgh = 200kg $\times$ 10m/s/s $\times$ 1m = 2000J.

35. The exact temperature depends on many factors. Softwoods catch fire at lower temperatures than dense, heavy hardwoods. The lower figure of 200°C (400°F) is quoted in Cote, A. & Bugbee, P. 1988. *Principles of Fire Protection*, National Fire Protection Association, New York.

36. Generally this doesn't happen: sporting stars all understand the importance of the 'follow-through', in which you keep your leg, arm or another part of your body moving after contact to increase the amount of energy you pass on, reduce the force on your limbs and lessen the risk of injury.

37. At school, hydraulic mechanisms like this are explained by a scientific law called Pascal's Principle, dating from 1651, which effectively says that the pressure of the water is the same at all points in a pipe. So, if you divide the force of the fluid by the area of the pipe (to give the pressure), you'll get the same number even if the pipe widens or narrows; a small force on a small pipe translates into a big force on a big pipe, in the case of the hydraulic jack. I think it makes more sense to try to understand hydraulic machines in terms of energy. The energy the jet has when it leaves a water pistol can't be any greater than the energy you put in by squeezing the trigger. Since the water leaving the pistol is travelling faster than the trigger, it moves through a bigger distance each second. The energy something has is equal to the force acting on it multiplied by the distance it moves. If the energy remains constant (as it must), the force must get smaller if the speed increases.

38. Cross, R. (2011). *Physics of Baseball and Softball*. Springer, New York.

39. Lance Armstrong & Oprah Winfrey: Cyclist Sorry for Doping. *BBC News*, 18 January 2013. See my website for the link.

40. For a fascinating insight into what top cyclists are capable of, see Padilla, S. *et al.* (2000), Scientific Approach to the 1-h Cycling World Record; *Journal of Applied Physiology*, 89, 1522–1527.

41. Cars typically weigh around 1500kg (3300lb), while bikes weigh about 15–20kg (33–44lb).

42. Alam, F. *et al.* 2009. Aerodynamics of Bicycle Helmets. In Estivalet, M. and Brisson, P. (2009), *The Engineering of Sport*. Springer, Paris.

43. Mack, J. 2007. Don't be a drag. *Bicycling*, August 2007, p. 46.

44. For a variety of plausible suggestions, see Sumner, J. 'Why Do Cyclists Shave Their Legs?' *Bicycling* (9 October 2014), at www.bicycling.com/. See my website for the link.

45. See Wilson, D. (2004), *Bicycling Science* (MIT Press, Cambridge, MA), p. 188.

46. Ward's Auto, the motor industry analysis centre, announced this in World Vehicle Population Tops 1 Billion Units. *Ward's Auto* (15 August 2011), via wardsauto.com/. See my website for the full link.

47. The figure of six billion is based on the number of SIM cards, not the number of users or handsets. UN: Six Billion Mobile Phone Subscriptions in the World, *BBC News*, 12 October 2012. See my website for the full link.

48. One billion sheep cited by Compassion in World Farming, *Factsheet: Sheep*, accessed 7 December 2013, via www.ciwf.org.uk. See my website for the full link.

49. At 3000 rpm, a 3.8-litre Porsche Turbo uses $(3000 \times 3.8)/2 = 5,700$ litres per minute. If a cyclist breaths about 10 times per minute at 50 per cent of lung capacity (50 per cent of 5 litres), that suggests they need only 25 litres per minute. Similar figures are quoted by Hammond, R. (2008), *Car Science* (Dorling Kindersley, London), p. 22.

50. American Motors Corporation offered a high-altitude package on its cars, which 'brought a special engine/gearbox/final drive combo, factory adjustments to carb fuel mixture, engine idle speed, and ignition timing and some changes to emission controls', according to Cranswick, M. (2011), *The*

Cars of American Motors (McFarland Publishers, Jefferson, NC), p. 187.

51. Denny, M. (2011). *Gliding for Gold: The Physics of Winter Sports*. Johns Hopkins University Press, Baltimore, MD. See 'Note 10: Getting High on Speed'.

52. A Ford Focus weighs in at about 1800–1900kg (4000–4200lb) according to Ford Focus: Features and Specifications. See www.ford.co.uk/Cars/Focus/Featuresandspecifications.

53. United States Office of Technology Assessment, Congress (1995). *Advanced Automotive Technology: Visions of a Super-efficient Family Car*. Diane Publishing, Darby, PA, p. 203.

54. Fuel Economy: Where the Energy Goes. *US Department of Energy Office of Energy Efficiency and Renewable Energy*. See www.fueleconomy.gov/feg/atv.shtml.

55. Professor Ferdinand Porsche Created the First Functional Hybrid Car. *Porsche*, 20 April 2011. See press.porsche.com/; the full link is on my website.

56. Electric car maker Tesla quotes about 40km (23 miles) of range per hour of charging, as a typical figure, although its superchargers promise a much more convenient 275km (170 miles) of range for just 30 minutes of charging. See Tesla Charging, at www.teslamotors.com/charging.

57. This chart was compiled by comparing data from three different sources: Energy Density via Wikipedia (see my website for link); Hammond, R. (2008), *Car Science* (Dorling Kindersley, London), p. 83; MacKay, D. (2008), *Sustainable Energy Without the Hot Air* (UIT Cambridge, Cambridge), p. 199, which gives figures in kWh that I've converted to MJ.

58. There's no exact figure for how hot a car's brakes get: it depends on all kinds of factors, including the type of material they're made from, the mass of that material, how fast the car is going, how well the brakes are cooled, the ambient temperature, whether it's raining and so on. For a ballpark figure, I consulted Burt, W. (2001), *Stock Car Race Shop* (Motorbooks International, Osceola, WI, p. 199), which says brakes 'begin to glow' at 650°C (1200°F). Stapleton, D. (2008), *The MG Midget and Austin Healey Sprite High Performance Manual* (Veloce Publishing, Dorchester, p. 124) suggests the upper limit for typical car brakes is 700°C

(1292°F). For extremes of speed, the official Formula 1 website quotes a figure of 750°C (1382°F). See 'Brakes', via www.formula1.com/. There is a full link on my website.

59. United States Office of Technology Assessment's *Advanced Automotive Technology: Visions of a Super-efficient Family Car* (Diane Publishing, Darby, PA, p. 165) suggests a typical value is 8–10 per cent, and 17–18 per cent is the maximum. See also Drivers on Track for Greener Trains (*BBC News*, 10 October 2009 – link on my website), which says 'train companies can save around 15% of their energy use' through regenerative braking.

60. Motorbikes are just as impressive: the contact area between tyre and road is a 'postage-stamp sized patch of material, which started its life in a tree'. See Dunlop (2012) Sportmax Range, via www.dunlop.eu/. There is a full link on my website.

61. Vollmer, M. & Möllmann, K. 2013. Is There a Maximum Size of Water Drops in Nature? *Physics Teacher*, 51, 400. A link to an online version is on my website.

62. Tiny, short-range electrostatic forces between nearby molecules are called Van der Waals forces, after a Dutch physicist named Johannes van der Waals (1837–1923). See Johannes Diderik van der Waals: Biographical: The Nobel Prize in Physics 1910, via nobelprize.org/nobel_prizes/physics/laureates/1910. There is a full link on my website.

63. There are 6×10^{23} molecules in 18g (0.6oz) of water, so there are 3×10^{21} molecules in 0.1g (0.004oz), which is quite a large drop of water. Glue would have much bigger and heavier molecules than water. Even so, we're in the right ballpark to consider that a single drop of glue would contain many trillions of molecules.

64. One drop can support about 23 newtons (2.3kg or 5lb), according to Science X Network of 12 December 2013, Nature's Strongest Glue Could be Used as a Medical Adhesive (see phys.org/ – full link on my website). The sticking power of *Caulobacter crescentus* is described in the same article.

65. Sticky Moments in 21 Years of Superglue. *BBC News*, 21 October 1998. See my website for the link.

66. Silver, S. *et al.* 1975. US Patent 3,922,464: Removable Pressure-Sensitive Adhesive Sheet Material. 25 November 1975. See my website for a link.

67. The particle size is 25–45 microns compared to a conventional adhesive, which is more like 0.1–2.0 microns, according to Karukstis, K. & Van Hecke, G. (2003), *Chemistry Connections: The Chemical Basis of Everyday Phenomena* (Academic Press, New York), p. 214.

68. 'Such forces become negligible at distances equivalent to only about 4 or 5 atomic diameters' according to Hecht, E. (1998), *Physics Algebra/Trig* (Brooks/Cole, Pacific Grove, CA), p. 114.

69. 'A billion hairs' comes from Ruibal, R. & Ernst, V. (1965). The Structure of the Digital Setae of Lizards (*Journal of Morphology* 117: 271–294). For a broader discussion of gecko glues, see Huber, G. *et al.* (2005), Evidence For Capillarity Contributions to Gecko Adhesion from Single Spatula Nanomechanical Measurements (*Proceedings of the National. Academy of Sciences*, 102, 16293–16296. A link to this paper is on my website.

70. A gecko weighing about 150g (5oz) can support a weight of 40kg (90lb), according to *Physics.org* (see www.physics.org/facts/gecko-really.asp), which means it supports 267 times its own body weight. So if a human weighing 75kg (165lb) scaled up in exactly the same way, they could carry about 20,000kg or 20 tonnes. For a more accurate calculation, we'd need to consider the relative size of a gecko's feet and a person's hands and feet and take into account the fact that less of a human's body would come into contact with the ceiling.

71. Slippery Slope: Researchers Take Advice from Carnivorous Plant. From Harvard School of Engineering and Applied Sciences (news, 21 September 2011), via www.seas.harvard.edu/news. See my website for the link.

72. For an interesting slant on the physics of ice sports such as curling, see sportsnscience.utah.edu/curling-friction-technical/.

73. You can watch Richard Feynman explaining the slipperiness of ice in an excerpt from the wonderful BBC series Fun to

Imagine at bit.ly/1wLEk9j. Note how he very neatly sidesteps responsibility for someone else's theory with the insertion of a strategic '*they say*'.

74. This is the theory expounded by Somorjai, G. & van Hove, M. in Getting a Grip on Ice. *Science*, 9 December 1996, via news.sciencemag.org/technology. See my website for the full link.

75. According to two answers on the Argonne National Laboratory 'Ask a Scientist' website, a typical cloud has a volume of several billion cubic metres. Assuming a liquid water content of about 0.3g (0.01oz) per cubic metre, we can calculate that the volume of the water in the cloud is probably somewhere between a few hundred thousand and a few million litres. Figures for cloud water content come from Linacre, E. & Geerts, B. Cloud Liquid Water Content, Drop Sizes, and Number of Droplets, www-das.uwyo.edu/~geerts/cwx/notes/chap08/moist_cloud.html. See my website for the link.

76. Cambridge University's Cavendish Laboratory has a clear and simple overview of J. J. Thomson's 1897 experiment. See Cambridge Physics: Discovery of the Electron, www-outreach.phy.cam.ac.uk/camphy/electron/electron_index.htm.

77. The anecdote is told by Thomson's grandson David in Davis, E. & Falconer, I. (1997), *J. J. Thomson and the Discovery of the Electron* (Taylor & Francis, London).

78. It was a Nobel sacrifice, as well as a noble one. Marie and her husband Pierre jointly won the 1903 Nobel Prize in Physics for this work, with Henri Becquerel. See The Nobel Prize in Physics 1903, via www.nobelprize.org/nobel_prizes/physics/laureates/1903/.

79. Movie of the Week: Madame Curie, *LIFE*, 13 December 1943, pp. 118–122.

80. Berkeley physics professor Richard Muller estimates that over a period of 50 years, the slightly higher natural radioactivity of Denver, Colorado will 'cause 4,800 excess cancer deaths. That's more excess death than is expected from the Chernobyl nuclear accident!'. See Muller, R. (2008), *Physics for Future Presidents* (W. W. Norton, New York), p. 117.

81. Gleason, S. (1955). Finding Uranium in the Dark. *Popular Science*, July 1955, p. 71.

82. Schoolboy, 13, Creates Nuclear Fusion in Penwortham. *BBC News*, 5 March 2014. See my website for the link.

83. Image of a 7 TeV proton–proton collision in CMS producing more than 100 charged particles from CERN, at cds.cern.ch/record/1293117. This artwork is copyright © CERN 2010 and reproduced here under CERN's copyright conditions allowing 'educational and informational use'. See my website for the link.

84. An excellent website produced by the Particle Data Group at Lawrence Berkeley National Laboratory (LBNL) is the place to start if you're new to atomic physics. See www.particleadventure.org/.

85. There are 6×10^{23} atoms in 235g (8oz) or one mole of uranium-235 and each splitting atom will release about 3.2×10^{-11}J according to Sowerby, Kaye & Laby, Tables of Physical and Chemical Constants, via www.kayelaby.npl.co.uk/ (see my website for the full link). So 235g of uranium-235 will produce about 20 trillion joules and 1g of uranium will produce around 100 gigajoules (or 100 gigawatts if it happens in one second).

86. This kind of atomic arrangement is called close-packing. Iron (a typical metal) has an interior, crystalline form called body-centred cubic (BCC), which means the atoms are arranged in the form of a cube, with eight sitting at the edges and one more squatting in the centre, in between the others. The basic 'unit cell', as this is called, repeats over and over again, like a kind of three-dimensional patterned wallpaper.

87. Bacteria digest trash in landfills, and there is some evidence that marine bacteria can pull off the same trick. See Zaikab, G. (2011), Marine Microbes Digest Plastic, from www.nature.com of 28 March 2011. See my website for the link..

88. Nylon, the first real synthetic plastic textile, was launched on 27 October 1938. There were earlier plastics, but the arrival of nylon marked the true beginnings of the modern plastic age.

89. The figures in this chart are, by necessity, a rough indication. They're compiled from various different sources including

Household Waste – That's Garbage!, Michigan Waste Stewardship Program, via www.miwaterstewardship.org/ (full link on my website); Save Our Beach, www. saveourbeach.org; and Surfers Against Sewage (2010), *Motivocean: Marine Litter: Your Guide* (Surfers Against Sewage, St Agnes, Cornwall).

90. Kevlar. See www.explainthatstuff.com/kevlar.html.

91. Professor Eric Le Bourhis dates the first glass to somewhere between 3500 and 5000 bc. See Le Bourhis, E. (2008), *Glass* (Wiley-VCH, Weinheim), p. 29.

92. See Fosbroke, T. (1843). *Encyclopaedia of Antiquities and Elements of Archaeology, Classical and Mediaeval, Volume 1.* M. A. Nattali, London. 'Beckman observes that transparent windows were in the time of Seneca quite novel, Stubbs ascribes the introduction here of stone and glass windows to Wulfrid, Bishop of Worcester, in 736'.

93. See the section 'Pigmenting glass' in Langhamer, A. (2003). *The Legend of Bohemian Glass: A Thousand Years of Glassmaking in the Heart of Europe* (Tigris, Czech Republic).

94. Szasz, F. 2006. 'J. Robert Oppenheimer and the State of New Mexico', in Kelly, C. (ed.) (2006), *Oppenheimer and the Manhattan Project.* World Scientific, Singapore.

95. For a very detailed discussion, see Zallen, R. (2008), *The Physics of Amorphous Solids* (John Wiley & Sons, Weinheim): 'In amorphous solids, long-range order is absent; the array of equilibrium atomic positions is strongly disordered'.

96. Debennetti, P. & Stanley, H. 2003. Supercooled and Glassy Water. *Physics Today*, June 2003, p. 40: 'Glassy water may be the most common form of water in the Universe. It is observed as a frost on interstellar dust, constitutes the bulk of matter in comets, and is thought to play an important role in phenomena associated with planetary activity'.

97. Szczepanowska, H. (2013). *Conservation of Cultural Heritage: Key Principles and Approaches.* Routledge, London. The section 'Myths about glass' quotes Dr Robert Brill, pointing out that 'the viscosity of glass is probably a billion times higher than metallic lead, and we never see lead flowing down from stained glass windows'.

98. In materials-science jargon, glass has a low *fracture toughness* and a low *work of fracture*. These are two related (but different) measurements of how much energy it takes to make a crack spread through a material. The incoming energy has to go somewhere, and if it can't deform a material, the material breaks instead. Balloons have such a low work of fracture that even the tiny point of a pin can make cracks spread through them instantly, making them burst violently.

99. See 'Coefficients of Cubical Expansion of Solids' in *Lange's Handbook of Chemistry* (1979; McGraw-Hill, New York).

100. The density of glass is about 2500 kg/m³, so a cubic metre of glass would weigh 2.5 tonnes.

101. Without going too deeply into the physics, this is essentially a band-gap argument, summarised in Smallman, R. & Bishop, R. (1999), *Modern Physical Metallurgy and Materials Engineering: Science, Process, Applications* (Butterworth-Heinemann, Oxford), p. 195.

102. McCollough, F. (2008). *Complete Guide to High Dynamic Range Digital Photography*. Lark Books, New York; p. 13 suggests that the average luminance in candela per square metre for a sunny sky is about 100,000, compared to just 50 for indoors.

103. See Boyd, R. (1957), Design of glass for daylighting. In *Windows and Glass in the Exterior of Buildings: A Research Correlation Conference Conducted by the Building Research Institute*. (National Academies Press, Washington), p. 8.

104. Ferrell McCollough (see note 102 above) quotes indoor light as 50 candela per square metre.

105. For a detailed explanation of electrochromic windows, see Arntz, F. *et al.* 1992. US Patent 5,171,413: Methods for Manufacturing Solid State Ionic Devices, 15 December 1992, (see my website for link). The essential similarity between the two technologies is obvious from the very first sentence, describing a 'device usable as an electrochromic window and/or as a rechargeable battery'.

106. Armistead, W. & Stookey, S. (1962), US Patent 3,208,860: Phototropic Material and Article Made Therefrom. 28 September 1965. See my website for the link.

107. 'In 1928, Walter Diemer … created bubblegum. He used a rubber tree latex'. See Mathews, J. (2009), *Chicle: The Chewing Gum of the Americas, From the Ancient Maya to William Wrigley* (University of Arizona Press, Tuczon, AZ).

108. 'Chewing gum contains a base that is made from natural rubber, styrene butadiene, or polyvinyl acetate'. Askeland, D. *et al.* (2010), *Essentials of Materials Science and Engineering.* Cengage Learning, Stamford, CT, p. 527.

109. If you've got a pair of polarising sunglasses (non-polarising ones won't work) and a laptop or tablet computer, you can experiment with photoelasticity for yourself. Open a word-processing program with a blank document and maximize it so the whole screen is completely white. Now put on your sunglasses and hold some transparent plastic objects between your eyes and the computer screen. You should see some amazing psychedelic spectral colours. What happens when you stress the objects or rotate your head?

110. Professor James Gordon argued that the invention of the pneumatic tyre was as significant as the internal combustion engine: '… if an effective pneumatic tyre had been available around 1830, we might have gone direct to mechanical road transport without passing through the intervening stage of railways at all'. Gordon, J. (1978), *Structures.* Penguin, London, p. 314.

111. The modulus of elasticity of steel (also called Young's modulus) is about 200,000MPa, compared to rubber, which is about 1MPa. Glaser, R. (2001), *Biophysics* (Springer, Berlin), p. 213.

112. Sun, J. *et al.* (2012). Highly stretchable and tough hydrogels. *Nature* 489, p. 133–136.

113. In *Structures* (see note 110 above), p. 54, James Gordon quotes a figure of 0.2MPa for the modulus of elasticity of a pregnant locust compared to 7MPa for rubber.

114. See Section 5.5 The Reversibility in Chandrasekaran, V. (2010), *Rubber as a Construction Material for Corrosion Protection.* John Wiley & Sons, Hoboken.

115. For a general discussion of the mechanical properties of facial skin, see Piérard, G. *et al.* (2010), Facial Skin Rheology. In Farage, M. *et al.* (eds), *Textbook of Aging Skin.* Springer, Heidelberg.

116. The modulus of elasticity of glass is about 70,000MPa, which suggests glass is at least twice as elastic as steel.

117. This figure is loosely inspired by the more abstract drawing of how cracks concentrate stress (p. 66) in Gordon, J. 1978. *Structures*. Penguin, London.

118. For a brief overview, see When Metals Tire, in Levy, M. & Salvadori, M. (1992), *Why Buildings Fall Down* (W. W. Norton, New York). There is a more detailed discussion in Withey, P. (1966), Fatigue Failure of the De Havilland Comet I., in Jones, D. (ed). (2001), *Failure Analysis Case Studies II* (Elsevier Science, Oxford).

119. Goldsmith-Carter, G. 1969. *Sailing Ships and Sailing Craft*. Hamlyn, London.

120. For a description of how the naturally flexible wooden hull of the *St Roch* enabled it to break free of the Arctic ice, see Delgado, J. (1985), *Across the Top of the World: The Quest for the Northwest Passage* (Douglas & McIntyre, Vancouver), p. 185.

121. The Polar Ship *Fram*, See www.frammuseum.no/. See my website for the full link.

122. See Chapter 4: Leather Preservation. In Smith, C. (2003), *Archaeological Conservation Using Polymers: Practical Applications for Organic Artifact Stabilization*. Texas A&M University Press, College Station.

123. For a more detailed explanation, see my article on thermochromic color-changing materials at www. explainthatstuff.com/thermochromic-materials.html.

124. Clout, L. Splendour of new Wembley fading already. *The Telegraph*, 10 May 2007.

125. Pohanish, R. 2011. *Sittig's Handbook of Toxic and Hazardous Substances*. Elsevier, Oxford, p. 736.

126. Johnson, J. 2011. *Old-Time Country Wisdom and Lore: 1000s of Traditional Skills for Simple Living*. Voyageur Press, Minneapolis, p. 51.

127. A figure of 5km/s (three miles per second) is quoted by Mia Siochi in NASA's educational video 'Real World: Self Healing Materials' on YouTube. See my website for the link.

128. Different sources say 1703 and 1704, but the earliest facsimiles of the cover I've found clearly show a date of MDCCIV (1704). Newton had first published his thoughts

about light some 30 years before (coincidentally, at the age of 30), in a paper in the *Philosophical Transactions* of 1672.

129. Thanks to Cambridge University, you can flick through the pages of Newton's notebooks from the comfort of your computer. See 'Isaac Newton: Laboratory Notebook' on the *Cambridge University Digital Library*, cudl.lib.cam.ac.uk. My website has the full link.

130. Wave-particle duality (the concept that light behaves as *both* particles and waves) is not a new idea. For a good account of how the pioneers of optics saw light in the 18th century, see Shamos, M. (1959). *Great Experiments in Physics* (Dover, New York), p. 93.

131. In 1900, Lord Kelvin (William Thompson) daringly told the British Association for the Advancement of Science that 'There is nothing new to be discovered in physics now'. Although this widely recycled quote has been disputed, it seems to be an accurate summary of what he believed. In a paper published the following year, he pointed out that the 'beauty and clearness' of physics was obscured only by two clouds – precisely the areas that would soon be explored by relativity and quantum theory. Kelvin, Lord (1901), Nineteenth Century Clouds over the Dynamical Theory of Heat and Light. *Philosophical Magazine and Journal of Science*, S. 6., 2. p. 1.

132. Hecht, E. 1998. *Physics: Algebra/Trig.* Brooks Cole, Pacific Grove, p. 806.

133. No-one seems to be able to put an exact figure on this, but there are credible estimates varying from 25–60 per cent in many books and across the Web. See my website for a selection.

134. Interestingly, Newton had a 'premonition' of Einstein's equation three hundred years earlier. In *Opticks*, question 30, he writes: 'The changing of Bodies into Light, and Light into Bodies, is very conformable to the Course of Nature.' – which sounds very much to me like $E = mc^2$.

135. In terms of length alone, the difference between the astronomical and the 'atomical' is vast. If we divide the estimated diameter of the observable universe (100 billion light years) by the diameter of an atom (0.25 nanometres), we get about 4 trillion trillion trillion (4×10^{36}) or, in old money, 4,000,000,000,000,000,000,000,000,000,000,000,000.

136. For a useful summary of the electromagnetic spectrum, see What Wavelength Goes With a Color? NASA, via science-edu.larc.nasa.gov. My website has the full link.

137. This is the average sort of speed reached by surfers, according to one of the classic books on ocean waves, Bascom, W. (1980), *Waves and Beaches*. Anchor Press, New York. Open ocean waves travel much faster, especially in tsunamis.

138. Interestingly, because light is so much bigger than atoms, we have no hope of seeing either atoms or molecules with ordinary optical (light-based) microscopes. That's why electron microscopes were invented. Since (in a crude and simplistic sense) electrons are much 'smaller' than photons, they can make images of much smaller things. The word 'smaller' becomes problematic when we talk about the probabilistic subatomic, but let's not worry about that here.

139. Helium-neon red laser photons each have about three-billionths of a joule (3×10^{-19}J) of energy according to Hecht, E. (1998), *Physics: Algebra/Trig* (Brooks Cole, Pacific Grove), p. 807.

140. A typical flashlight bulb is about 2.2 volts and 0.25 amps, which makes 0.55 watts. Dividing 0.55 watts by the energy of a photon quoted above (3×10^{-19} J) gives us about 2×10^{18} photons per second.

141. Cathcart, B. (2005). *The Fly in the Cathedral*. Farrar, Straus and Giroux, New York.

142. Although energy zaps from the Sun to Earth in just a few minutes, it takes *thousands of years* for it to get from the nuclear core of the Sun to the outer layers so it can make its escape. See Plait, P. (1997), The Long Climb from the Sun's Core, via www.badastronomy.com/bitesize/solar_system/.

143. Molten volcanic lava comes in at about 1,200°C (2,200°F) according to Schminke, H. (2004). *Volcanism* (Springer, Berlin), p. 27.

144. For an entertaining account of the extent to which the idealised myth of Thomas Edison departs from what we can establish of historical reality, see Stross, R. (2007), *The Wizard of Menlo Park* (Crown, New York). The infamous guard bear is described in Jehl, F. (1924), Edison the Man: An Old Friend's Recollections of the Great Inventor. *Popular Science*, February 1924, p. 31.

145. Silicon, one of the commonest chemical elements on Earth, is often confused with silicone, a type of rubbery plastic (polymer) used in breast implants. Although silicone contains silicon, that's where the resemblance begins and ends. The silicon used in 'silicon chips' (computer microchips) is much closer to the raw stuff you find in sand than the stuff lurking in a brassiere.

146. Sean Palmer suggests a firefly produces light of about 1/50 and 1/400 candlepower. See 'What distance is a firefly visible from?' *in Sean B. Palmer's Shared Objects*, sbp.so/firefly.

147. For a wonderful account of the Hubble's troubles, see Zimmerman, R. (2010), 'Chapter 4: Building it' and 'Chapter 5: Saving it' in *The Universe in a Mirror: The Saga of the Hubble Telescope and the Visionaries who Built it* (Princeton University Press, Princeton, NJ).

148. The 'luminiferous aether' (sometimes spelled ether) finally packed its bags after the famous 1887 Michelson-Morley experiment, in which Albert Michelson and Edward Morley devised a cunning test to detect its existence and measure its speed. Finding a speed of zero, they effectively proved the aether didn't exist, suggested the speed of light was always the same – and paved the way for Einstein's shiny new theory of relativity about 20 years later.

149. A Chat with the Man Behind Mobiles. *BBC News*, 21 April, 2003 (see my website for link). Martin Cooper's main patent for the mobile phone was filed in 1973 and granted two years later. See Cooper, M. (1975), US Patent 3,906,166: Radio Telephone System, 16 September 1975 (link available on my website).

150. In an entertaining history of how the UK and North America were linked up electrically, Gillian Cookson suggests it took about 12 days to send a message across the Atlantic by a combination of steam ship and electric telegraph during the 1850s. See Cookson, G. (2012), *The Cable* (The History Press, Stroud). Once Cyrus Field's pioneering submarine cable between Ireland and

Newfoundland was completed in 1858, it was soon sending about 150 messages a day (the total traffic in both directions). A PBS article about Field's cable suggests it was sending only 50 messages a day in 1866, mostly due to the high cost of transmission ($10 per word). See The Great Transatlantic Cable, www.pbs.org/wgbh/amex/cable/index.html.

151. It takes light travelling at 300,000km/s (186,000 miles/s) about 0.02 seconds to travel the straight-line distance of approximately 5,500km (3,400 miles) between London and New York City.

152. See, for example, the classic paper Nyquist, H. (1924), Certain Factors Affecting Telegraph Speed (*Bell System Technical Journal*, 3, 324–346). See my website for a link.

153. For a surprisingly entertaining account, see the Elisha Gray and Alexander Bell Telephone Controversy on *Wikipedia* (see my website for a link). Few people had heard of Meucci until his contribution was finally recognized in a resolution passed by the United States House of Representatives on June 11, 2002. See Who is Credited as Inventing the Telephone? via the US Library of Congress, www.loc.gov/rr/scitech/mysteries/telephone.html.

154. The photophone is described in detail in Bell, A. (1880), US Patent 235,199: Apparatus for Signalling and Communicating, called 'Photophone', 7 December 1880. See my website for a link.

155. *TIME* magazine went some way to setting the record straight by naming Farnsworth one of its 100 most influential figures of the 20th century. See Postman, N. (1999), Electrical Engineer Philo Farnsworth, *TIME*, 29 March 1999. See my website for a link.

156. Suppose you're listening to FM radio on a frequency of about 90MHz. The wavelength is the speed of light divided by the frequency or about 3.3m (11ft), so a suitable FM antenna would be round about 1.5m (5ft) long.

157. The maximum distance you can see a lighthouse is 3.57km (2.2 miles) multiplied by the square root of its height in metres. So, if a lighthouse is about 30m (100ft) tall, you can see it about 20km (12 miles) away. I'm using a formula by Young,

A. Distance to the Horizon at mintaka.sdsu.edu/GF/explain/ atmos_refr/horizon.html. There's a link on my website.

158. Marconi won the 1909 Nobel Prize in Physics. The lecture he delivered is a fascinating, first-hand account of his experiments and the scientific understanding of the time; the transcript is well worth reading. Marconi, G. (1909), Nobel Lecture: Wireless Telegraphic Communication, via www.nobelprize.org/. See my website for the link

159. It was simultaneously developed by Arthur Kennelly and finally confirmed by Edward Appleton in 1924. Appleton won the 1947 Nobel Prize in Physics for this work – an honour that Kennelly and Heaviside, both then deceased, were unable to share. See Edward V. Appleton – Biographical, on www.nobelprize.org/ (full link on my website).

160. These and other colourful anecdotes are quoted by Nahin, P. (2002). *Oliver Heaviside: The Life, Work, and Times of an Electrical Genius of the Victorian Age* (JHU Press, Baltimore, MD).

161. Searle, G. 1950. Oliver Heaviside: A Personal Sketch. In *The Heaviside Century Volume* (IEE, London).

162. Clarke, A. 1945. Extra-Terrestrial Relays: Can Rocket Stations Give World-Wide Radio Coverage? *Wireless World*, October 1945.

163. The Straight Dope website quantifies it as 1,000 watts for a microwave oven compared to a few hundred milliwatts for a cellphone. See How are the Microwaves in Ovens Different from those in Cell Phones? via The Straight Dope, www.straightdope.com, 28 August 2003. My website has the full link.

164. This is a wild guesstimate simply based on The Straight Dope's comparison of the power, assuming a mobile phone could beam microwave energy into food in exactly the same way as a microwave oven. In reality, a cellphone wouldn't generate anything like enough *temperature* to cook food: even if it could eventually beam enough energy into your dinner, that's not the same as cooking it. You can leave your dinner outside in the Sun for as long as you like, but it still won't cook unless it reaches a certain, critical temperature.

165. ENIAC: Celebrating Penn Engineering History. Penn Engineering, via www.seas.upenn.edu. My website has the full link.

166. Van der Spiegel, J. (2001) ENIAC. In Rojas, R. (ed.), *Encyclopedia of Computers and Computer History* (Fitzroy Dearborn, Chicago).

167. See for example the article DigitalVersus Film Photography on Wikipedia. See my website for a link.

168. The only feasible way to listen in on digital mobile calls is to install listening equipment in the handset or tap into the equipment in the phone provider's switching office. That's why hacking into mobile phone messages became such a popular pastime for journalists in the period from the mid-1990s to the mid-2000s. An old telecoms industry friend of mine, who knows about such things, once told me the mobile phone companies have sinister, high-security rooms fitted with secret snooping equipment to which only the security services have access – an idea I instantly brushed aside as conspiracy-theory paranoia. In 2014, the mobile phone operatorVodafone finally confirmed that it is indeed true. See Garside, J. 2014.Vodafone Reveals Existence of Secret Wires that Allow State Surveillance (*The Guardian*, 6 June 2014).

169. The Library of Congress has an ever-expanding collection: 36 million books is the figure cited at the time of writing. See Fascinating Facts from the US Library of Congress, loc. gov/about/facts.html.

170. iTunes began dropping DRM from its music files in 2009. See Apple to End Music Restrictions, *BBC News*, 7 January 2009. Full link on my website.

171. In the United States, at least, under the Digital Millennium Copyright Act (DMCA), signed into law by President Bill Clinton on 28 October 1998.

172. This method of cracking copyright protection is sometimes referred to as the Analog(ue) Hole – and the media multinationals have their sights set on trying to close it. See the Electronic Frontier Foundation, www.eff.org/issues/analog-hole.

173. In theory, if you sample fast enough, you could capture all the important information in your original sound. We know this from something called the Nyquist-Shannon sampling theorem. In practice, sampling too fast makes MP3 files too big and most people (those who listen to bad MP3s through cheap headphones on noisy commuter trains) are happy to tolerate poorer-quality, highly compressed files if it means they can store far more music on their iPod or mobile phone.

174. CDs are much more impressive than I've made them sound – and seemed even more so when they were first introduced. If you're old enough, you might remember TV demonstrations from the early 1980s where hapless science presenters dropped CDs on the floor or smeared them with jam to prove that they were still playable after rough treatment. That kind of thing is made possible by a handy bit of mathematical magic built into their design, called the cross-interleave Reed–Solomon error-correction code (CIRC), which allows them to whistle past random errors caused by things like scratches and fingermarks. It's so good that it can (theoretically – and very impressively) compensate for a scratch 4,000 bits or 2.5mm (0.1in) long. See Baert, L. (1995), *Digital Audio and Compact Disc Technology* (Focal Press, Boston).

175. Thanks to the British Library, the Gutenberg Bible has been digitally preserved in perpetuity, but who knows if the digital facsimile will last as long as the paper original? See www.bl.uk/treasures/gutenberg/background.html.

176. Ray Tomlinson has explained the thinking behind the first email. Why did he do it? 'Mostly because it seemed like a neat idea'. Why the strange at sign @? 'The primary reason was that it made sense. At signs didn't appear in names so there would be no ambiguity about where the separation between login name and host name occurred.' See openmap.bbn.com/~tomlinso/ray/firstemailframe.html.

177. There's more about the BBC project at www.bbc.co.uk/history/domesday/story.

178. Estimated in a Radicati Group survey in 2012.

179. The weight of a typical iceberg is about 150,000 tonnes (165,000 tons) and its internal temperature is about $-15°C$ ($5°F$). So the heat energy it contains (heating it from absolute zero) is equal to its mass $\times$ specific heat capacity

of water $\times$ its temperature above absolute zero = 150,000,000kg $\times$ 4181J/kg/°C $\times$ (273–15) = 162,000 gigajoules. A very large cup of coffee might be about 0.5 litres and at a temperature of 90°C (194°F), so its heat energy is 0.5kg $\times$ 4181J/kg/°C $\times$ (90 + 273)°C = 760 kilojoules. The iceberg has about 200 million times more heat energy. My iceberg statistics come from www.canadiangeographic.ca/magazine/MA06/indepth/justthefacts.asp.

180. Black holes don't have to be cold, however, and space scientists are currently working to make 'the coolest spot in the Universe' (with a temperature of just 1 picokelvin, 10^{-12} °K or 0.000000000001°K). See coldatomlab.jpl.nasa.gov/.

181. Hand, E. (2012). Hot stuff: CERN physicists Create Record-Breaking Subatomic Soup. blogs.nature.com 13 August 2012. See my website for a link.

182. Bingelli, C. (2009). *Building Systems for Interior Designers*. John Wiley & Sons, Hoboken, NJ, p. 22.

183. Ice import: see www.canalmuseum.org.uk/ice/iceimport.htm.

184. Mrs. Marshall's Liquid Air Ice Cream. See matthew-rowley.blogspot.co.uk, September 2010. The full link is on my website.

185. I'm using figures for typical costs and savings compared to electric storage heaters calculated by the UK's Energy Saving Trust. See www.energysavingtrust.org.uk/Generating-energy/Choosing-a-renewable-technology/Ground-source-heat-pumps.

186. Quoted in Fox, S. (2010). Superinsulating Aerogels Arrive on Home Insulation Market At Last. See www.popsci.com/technology/article/2010-02/aerogels-hit-consumer-insulation-market.

187. The Passivhaus Standard: What is Passivhaus?, via www.passivhaus.org.uk/. See my website for the link.

188. See for example Fischer, B. How much does it cost to charge an iPhone 5? A thought-provokingly modest $0.41/year. On blog.opower.com, 27 September 2012. My website has the full link.

189. 'Without Much Fanfare, Apple Has Sold Its 500 Millionth iPhone', via www.forbes.com, 25 March 2014, See my website for the link.

190. Greenpeace International (2012). How Clean is Your Cloud? See www.greenpeace.org/international; my webpage has the full link.

191. Desktop versus laptop. See www.eu-energystar.org/en/en_022.shtml.

192. As explained later in this chapter, I am adopting the convention of using a capital to represent nutritional Calories, where 1 nutritional Calorie = 1000 thermal calories = 4.2 kilojoules (4200 joules) of energy.

193. The data for this pie chart is compiled from a number of different sources, including Insel, P. *et al.* (2010), *Nutrition* (Jones & Bartlett, Sudbury, MA), p. 342.

194. 'On average, efficiency ranges between 20 and 25% for walking, running, and stationary cycling.' McArdle, W. *et al.* (2010), *Exercise Physiology: Nutrition, Energy, and Human Performance* (Lippincott Williams & Wilkins, Baltimore, MD), p. 208.

195. Walking briskly for an hour at about 6km/h (4mph) consumes about 450 Calories. A large egg contains 75 Calories, so you could walk for about 10 minutes on an egg and go about 1km (0.6 miles).

196. The winter energy consumption of a musk oxen is about 920kJ (219 Calories)/kg of body weight, while summer consumption (when the animals are much more active) rises by about a quarter to 1,163kJ (278 Calories)/kg. See Kazmin, V. D. & Abaturov, B. D. (2011), Quantitative characteristics of nutrition in free-ranging reindeer (*Rangifer tarandus*) and musk oxen (*Ovibos moschatus*) on Wrangel Island (*Biology Bulletin*, 38, 935).

197. A hummingbird has a BMR to mass ratio of about 870kJ/hr per kg of body mass, while a human clocks in at just 77kJ/hr per kg, according to Sherwood, L. *et al.* (2012), *Animal Physiology: From Genes to Organisms* (Cengage, Independence, KY), p. 720.

198. Column 1 is compiled mostly using nutritional data from Gebhardt, S. & Thomas, R. (2002), Nutritive Value of Foods (*US Department of Agriculture Agricultural Research Service, Home and Garden Bulletin* 72, October 2002). A link is on my website. A few entries have been compiled directly from the nutritional labels on food from typical grocery store products.

199. My rough exercise equivalents are based *very* loosely on the figures suggested by Insel, P. *et al.* (2010) in *Nutrition* (Jones & Bartlett, Sudbury, MA), p. 344.

200. Mice eat 12g (0.4oz) of food per 100g (3.5oz) body weight per day according to the Johns Hopkins University Animal Care and Use Committee, web.jhu.edu/animalcare/procedures/mouse.html.

201. 1 litre of petrol contains 34.8 megajoules (34,800 kilojoules) of energy so 1 UK gallon (about 5 litres) contains about 160,000 kilojoules.

202. At the time of writing, and using UK prices, a litre of petrol costs about £1.30, which works out at about 0.00374p per kilojoule. A typical fast-food burger might cost £4.00, which is more like 0.2p per kilojoule (over 50 times more for the same energy). My cost for electricity is based on a unit price of 20p per kWh that doesn't take account of the annual standing charge. That works out at 0.006p per kilojoule.

203. Estimated Energy Requirements from Health Canada, 8 November 2011. See www.hc-sc.gc.ca; the full link is available on my website.

204. Breiter, M. *Bears: A Year in the Life*. A & C Black, London, p. 157.

205. Percent of Consumer Expenditures Spent on Food, Alcoholic Beverages, and Tobacco that were Consumed at Home, by Selected Countries, 2012. US Department of Agriculture Economic Research Service, www.ers.usda.gov/data-products/food-expenditures.aspx.

206. Wrangham, R. (2010). *Catching Fire: How Cooking Made us Human*. Basic Books, New York.

207. See www.ift.org/newsroom/news-releases/2013/july/15/chew-more-to-retain-more-energy.aspx.

208. Wolke, R. (2008), *What Einstein Told His Cook: Kitchen Science Explained*. W.W. Norton, New York. Or see McGee, H. (2004), *On Food and Cooking: The Science and Lore of the Kitchen* (Scribner, New York).

209. Unless, of course, we can find a way of making and digesting nuclear food. Einstein's equation $E=mc^2$ allows a pill weighing 1.5g to produce a theoretical 135 trillion joules of energy.

210. Dust ranges widely in size, but the smaller it is, the more likely we are to breathe it in, and the bigger the danger it poses. Typical dust ranges from a few microns (millionths of a metre) in diameter to a hundred microns or more.

211. This is correct in theory; in practice, the atmosphere is much more variable and complex and with phenomena such as inversion layers, the wind speed can actually decrease with height.

212. In theory, if you double the hub height (the height of the point around which the blades spin), the wind speed you tap into is about 10 per cent greater because (as a rule of thumb) wind speed increases with altitude to the power of 1/7, and that produces an increase in power of about a third. In practice, it's more complex. As the turbine rotates, each blade experiences higher wind speeds when it's at its highest point than when it's at its lowest point, so the load on the turbine increases too. Not only that, but bigger and taller turbines weigh more, so they lose energy by doing more work against gravity.

213. Swimming in the ocean doesn't really compare with swimming in a pool, for all sorts of reasons. Waves make it more turbulent and quite a bit of your energy is wasted tumbling through them. You might be more inclined to keep your head above the water in a cold, rough sea, so you'll adopt a much less streamlined posture and, consequently, create more drag that makes every stroke harder work. Swimming in a cold sea means you need to wear a wetsuit, gloves, boots and probably a hat – all of which keep you warmer but make it harder to move. Also, salty seawater is very slightly more dense than freshwater, and colder water is more dense than warm water; the added density of cold seawater (compared to warm pool-water) doesn't help, but it's much less important than the fluid dynamic factors.

214. NASA research in the 1970s found a 24 per cent reduction in drag at 90 km/h (55mph), according to Lamm, M. (1977; *Popular Mechanics*, p. 81). Fairings that improve drag around wheels are described in Rahim, S. (2011) in Plastic Fairings Could Cut Truck Fuel Use (*Scientific American*, February

2011; see www.scientificamerican.com/article/plastic-fairings-cut-truck-fuel/.

215. Einstein, A. (1926). The Cause of the Formation of Meanders in the Courses of Rivers and of the So-Called Baer's Law. *Die Naturwissenschaften*, 14.

216. In some sufferers the pharynx stays shut, producing a terrifying condition known as obstructive sleep apnoea (OSA). It can be treated with a device called a continuous positive airway pressure (CPAP) machine, which gently pumps air in through a face mask to keep your pharynx open. See the National Heart, Lung and Blood Institute's www.nhlbi.nih.gov/health/health-topics/topics/cpap/.

217. Fajdiga, I. 2005. Snoring Imaging – Could Bernoulli Explain It All? *CHEST*, 128, 896. See my website for a link.

218. The Water in You. US Geological Survey; see water.usgs. gov/edu/propertyyou.html.

219. A brilliant example of wave science that goes by the technical name of thin-film interference. You can read all about it at www.explainthatstuff.com/thin-film-interference.html.

220. Cavendish is often credited with discovering the composition of water, though James Watt claimed the same thing. There's an interesting discussion in Miller, D. (2004), *Discovering Water: James Watt, Henry Cavendish, and the Nineteenth Century 'Water Controversy'* (Ashgate Publishing, Farnham).

221. Indoor Water Use in the United States. US Environment Protection Agency (EPA), www.epa.gov/watersense/pubs/indoor.html.

222. Toilet flushing figures come from the US EPA web page in note 221. Figures for the Siemens iQ300/iQ500 clothes washing machine are 7 litres (2 gallons) of water per kilogram of dry laundry and a load capacity of 5–7kg (11–15lb). The link to Siemens Washing Machines is on my website.

223. Terry, N. (2011). *Energy and Carbon Emissions: The Way We Live Today*. UIT Cambridge, Cambridge, p. 47.

224. This is a variation on a classic physics exam question known as Caesar's dying breath, in which you're asked to calculate the likelihood of a breath of fresh air containing at least one molecule from the last lungful of the doomed Roman dictator. No calculation is required: all you really need to know is that there are more molecules in a glass of water (6 00,000,000,000,000,000,000,000 molecules in every 18g or 0.6oz of it, which is called a mole) than there are glasses of water on our planet (or in a lungful of air than there are lungfuls in Earth's atmosphere). Like the water, this factoid has been endlessly recycled. You'll find it in Dawkins, R. (2008), *The God Delusion* (Houghton Mifflin Harcourt, New York, p. 410), which itself recycles an earlier discussion by Lewis Wolpert noting that 'there are many more molecules in a glass of water than there are glasses of water in the sea'. I'm simply doing my duty by recycling it again.

225. The specific heat capacity of iron is about 0.47 kilojoules per kilogram per degree, or roughly nine times less than the specific heat capacity of water. That means it takes nine times more energy to heat up 1kg (2.2lb) of water by a certain temperature than 1kg of iron (or that you get nine times the temperature rise, if you prefer).

226. For the sake of a simple explanation, I've described home central heating as though it's a 'series' circuit with each radiator fed in turn from the boiler – which is how older systems were often fitted. In practice, newer systems have much more efficiently designed 'parallel' (trunk and branch) circuits with the water splitting up and taking different paths to different radiators instead of visiting each radiator in turn.

227. Operating temperatures for nichrome (nickel, chromium alloy) heating elements are about 750°C (1380°F).

228. Lifting the Lid on Computer Filth, *BBC News*, 12 March 2004; and Keyboards 'Dirtier Than a Toilet, *BBC News*, 1 May 2008. My website has the links.

229. See note 223.

230. World Health Organization (2003). *Guidelines for Safe Recreational Water Environments Volume 1: Coastal and Fresh Waters.* WHO, Geneva, p. 45.

231. Botcharova, M. 2013. A Gripping Tale: Scientists Claim to Have Discovered Why Skin Wrinkles in Water. *The Guardian*, 10 January 2013. See my website for a link.

232. Dyson Airblade Technical Specification: AB14, via www. dysonairblade.co.uk/hand-dryers/airblade-db/ airblade-db/tech-spec.aspx

233. Bakalar, N. (2003). *Where the Germs Are: A Scientific Safari.* John Wiley, New York, p. 54.

234. Handwashing: Why are the British so Bad at Washing their Hands?, *BBC News*, 15 October 2012. See my website for the link.

235. Most People Washing their Hands. From Guinness World Records; see www.guinnessworldrecords.com/. My website has the full link.

236. Quoted by Nicholas Bakalar (see note 233), p. 53.

237. US figures for 1997 quoted in Carpenter, R. (1999). Laundry Detergents in the Americas: Change and Innovation as the Drivers for Growth; in Cahn, A. (ed.) (1999), *Proceedings of the 4th World Conference on Detergents: Strategies for the 21st Century* (AOCS Press, Urbana, IL). European figure is quoted in Garratt, B. (2010). *The Fish Rots From The Head* (Profile Books, London).

238. Wilson, E. O. (1992). *The Diversity of Life.* Harvard University Press, Cambridge, MA, p. 142.

239. Estimates vary depending on the breed, but 50 million seems a good, average figure, according to Cook, J. (1984), *Handbook of Textile Fibres: Volume 1: Natural Fibres* (Woodhead Publishing, Cambridge UK), p. 89.

240. The idea of a 'universal solvent' dates back to the alchemists, who sought (in vain) for a substance they termed the Alkahest, which would dissolve everything else. Water is still the closest thing we have. See Chapter 1 in Reichardt, C. & Welton, T. (2011), *Solvents and Solvent Effects in Organic Chemistry* (Wiley-VCH, Weinheim).

241. Assuming each home uses 5,000 litres per year, you'd need 200 homes to use 1 million litres. A few hundred homes would give you several million litres, which is roughly the volume of an Olympic-sized swimming pool.

242. Say the diameter of the drum is 55cm (22in), giving a radius of 0.225m (11in) and a circumference of 1.41m (56in). If the drum rotates at 1000rpm, the circumference moves 1,410m (4,624ft) per minute = 85km/h (approximately 50mph).

243. MacKay, D. (2008). *Sustainable Energy Without the Hot Air.* UIT Press, Cambridge, p. 54.

244. Most people would say that the water is removed by centrifugal (literally, 'centre-fleeing') force, but science teachers don't like that term and would prefer to explain things a different way. They'd say the washing machine drum provides centripetal (literally 'centre-seeking') force that keeps the clothes moving around in a circle. Because there are holes in the drum, there's nothing to give the water inside the clothes centripetal force so it moves in a straight line, separates from the clothes, and escapes.

245. I tested this out on a cold December day a few years ago, when there was snow on the ground across much of the UK. I weighed a load of laundry before I hung it outside, left it blowing in the bitter easterly wind for about five hours, then brought it in and weighed it again. After washing and before any drying, it weighed 5kg (11lb); after outside drying, it clocked in at just 3.5kg (7.7lb). Then I dried it fully, indoors, and weighed it a third time. This time, the scales registered 3kg (6.6lb). Assuming that this was the dry weight, you can see that the outdoor drying removed about 75 per cent of the water.

246. See www.xeroscleaning.com/polymer-bead-cleaning/cost-benefits/.

247. The US EPA has a useful summary of the ingredients in a typical laundry detergent and their environmental impacts. See www.epa.gov/dfe/pubs/laundry/techfact/keychar.htm.

248. Endocrine disruptors, present in many chemicals (not just detergents), cause a significant proportion of fish to change sex. The US Geological Survey Toxic Substances Hydrology Program (4 August 2009; see my website for the link), quotes 18–22 per cent; Pollution changes sex of fish (BBC News, 10 July 2004, see my website for the link) quotes 'a third of male fish'; Pollutants in D.C. area drinking water in *The Washington Times*, 12 November 2009, quotes '80 per cent of fish' showing sex changes in Washington's Potomac River. For a much fuller discussion, see Kime, D. (1998), *Endocrine Disruption in Fish* (Kluwer Academic Publishers, Norwell, MA).

249. Simpson, W. S. & Crawshaw, G. H. (eds) (2002). *Wool: Science and Technology*. Woodhead Publishing, Cambridge, p. 303.

250. This is called Newton's Law of Cooling. One of its most interesting applications is in helping crime scene pathologists

to work out an approximate time of death from the temperature of a dead body.

251. Plowman, S. & Smith, D. (2007). *Exercise Physiology for Health, Fitness, and Performance*. Lippincott Williams & Wilkins, Baltimore, MD, p. 418.

252. The formal scientific name for this is heat of sorption. For much more about the science of wool, see the excellent booklet Leeder, J. (1984), *Wool: Nature's Wonder Fibre* (Australasian Textiles Publishers, Ocean Grove, Victoria).

253. Stainless steel comes in at 170–1000 MPa; wool manages 70–115 MPa. Data from Ashby, M. (2012), *Materials and the Environment: Eco-informed Material Choice* (Elsevier, Waltham, MA), pp. 466–592.

254. See www.gore-tex.com/remote/Satellite/home. The GORE-TEX website quotes the membrane as being '20,000 times smaller than a drop of water, but 700 times bigger than a molecule of moisture vapour'. You might wonder how these figures square with my estimate of a water drop having millions of trillions of molecules? As far as we can tell, Gore's scientists arrived at these figures by comparing *linear* dimensions rather than areas or volumes. It's important to remember that water drops vary hugely in size. I quickly ran the numbers for this book and arrived at a range of estimates from about 130 trillion up to several billion trillion water molecules in a drop.

255. Apart from discussing the fascinating similarities between structural materials and clothes, James Gordon shows how metal sheets can shear along a diagonal in exactly the same way as dresses and tablecloths, which is why you sometimes see a 'crease' in the fuselage of a jet plane or a helicopter. See Gordon, J. (1978), *Structures* (Penguin, London), pp. 248–259. As he notes, the bias cut was pioneered in the 1920s by French fashion designer Madeleine Vionnet.

256. Energy Expenditure During Walking, Jogging, Running, and Swimming. In McArdle, W. *et al.* (2010), *Exercise Physiology: Nutrition, Energy, and Human Performance*. (Lippincott Williams & Wilkins, Baltimore, MD), p. 210.

257. For a good introduction, see www.training-conditioning.com/2009/05/29/opening_the_gait/index.php.

Further Reading

In this little tour of the science hidden in your home, I hope I've demonstrated that science isn't a set of dusty old facts or meaningless equations – the way it's too often taught at school – but a fascinating way of looking at the world: a way of seeing the weird things around us so that they snap together in a new kind of sense. Thinking scientifically is like putting on glasses for the first time and marvelling at your crisply corrected eyesight.

If you've enjoyed reading this book, you might like to check out some of the following resources, similarly designed for non-scientists, which will take you further in your exploration of the science behind everyday life.

Websites

Explain that Stuff by Chris Woodford. My science-education website (www.explainthatstuff.com) covers some of the topics in this book in quite a bit more detail. There's an A–Z index and a guide to using the site for home learning and school study.

Books

Physics for Future Presidents by Richard Muller (W. W. Norton, 2008). Explores the science behind topics critical to our survival, including energy supply, climate change and terrorism.

Sustainable Energy Without the Hot Air by David MacKay (UIT Cambridge, 2009). How can we produce enough energy to satisfy our ever-increasing needs without wrecking the planet? MacKay uses razor-sharp reasoning, back-of-envelope calculations and school-level science to reach stark conclusions anyone can understand. The complete text is also available online at www.withouthotair.com.

Energy and Carbon Emissions: The Way We Live Today by Nicola Terry (UIT Cambridge, 2011). Does a shower use more energy than a bath? Is it better to heat your home with natural gas or electricity? This no-nonsense book settles arguments about how to live ethically with useful facts and figures.

The New Science of Strong Materials: Or Why You Don't Fall Through the Floor by J. E. Gordon (Penguin, 1991). A great introduction to how humble everyday materials like wood and glass hold our world together. James Gordon's other book *Structures* (Da Capo Press, 2009) is also worth looking out for.

Why Buildings Fall Down by Matthys Levy and Mario Salvadori (W. W. Norton, 2002). A set of fascinating case-studies describing how skyscrapers, stadiums, bridges and other structures have failed through basic problems of materials science and poor engineering.

Six Easy Pieces by Richard Feynman (Penguin, 1998). Feynman, one of the world's greatest physicists, was also one of its most inspiring science teachers. This short volume is a simple introduction to fundamental concepts like atoms, energy and gravity, although it lacks the off-the-cuff zest of his lectures and talks.

Mr Tompkins in Paperback by George Gamow (Cambridge University Press, 2005). What would happen if people shrank down to the size of atoms? An amusing introduction to the key concepts of atomic physics, first written in 1940 and just as valuable today.

Inflight Science: A Guide to the World from Your Airplane Window by Brian Clegg (Icon Books, 2011). A thoroughly enjoyable armchair journey through the science you'll encounter during a plane trip, explained with crystal clarity.

Why Things Are the Way They Are by B. S. Chandrasekhar (Cambridge University Press, 1997). Although somewhat more complex than the other books listed here, this is a wonderfully lucid look at solid-state physics – and the 'inside' explanations behind everyday things like magnetism and mirrors.

Acknowledgements

I owe an immense debt of gratitude to the following, without whom you wouldn't be reading this book now.

Jim Martin, champion of popular science, for setting up the Sigma list and kindly inviting me to join it. Thanks also to everyone at Bloomsbury who helped to put the book together and promote it.

Dr Jon Woodcock, for the rocket science that escaped me and many thoughtful suggestions that improved the book beyond measure.

Andrew Lownie, my persistently wonderful literary agent, without whom the book would never have been completed, much less published.

Index